The Analog Rebellion

Why We Are Ditching Screens and Loving the Digital Detox - How Vinyl, Film Cameras, and Flip Phones Are Making a Comeback in the TikTok Era

Les Brady

This publication is designed to provide accurate and authoritative information in regard to the subject matter covered. It is sold with the understanding that neither the author nor the publisher is engaged in rendering legal, investment, accounting or other professional services. While the publisher and author have used their best efforts in preparing this book, they make no representations or warranties with respect to the accuracy or completeness of the contents of this book and specifically disclaim any implied warranties of merchantability or fitness for a particular purpose. No warranty may be created or extended by sales representatives or written sales materials. The advice and strategies contained herein may not be suitable for your situation. You should consult with a professional when appropriate. Neither the publisher nor the author shall be liable for any loss of profit or any other commercial damages, including but not limited to special, incidental, consequential, personal, or other damages.

Contents

Introduction

The Analog Rebellion: A Manifesto for the Overwhelmed

Let's be real—our digital lives are a chaotic, notification-filled mess. At some point, we went from using the internet to check the weather and send the occasional email to having our every waking moment dictated by algorithms, endless scrolling, and the creeping suspicion that we're being watched (because, well, we are).

Once upon a time, we believed technology would free us—turns out, it just put us on an even shorter leash. Our pockets buzz with updates, our inboxes are bottomless pits, and our ability to focus is shot. Somewhere between the promise of "connection" and the reality of doomscrolling, we lost something fundamental: *the joy of slowness, the power of presence, and the satisfaction of doing things with our own two hands.*

That's why people are fighting back. The rise of vinyl records, film photography, typewriters, dumbphones, and paper planners isn't just nostalgia—it's a full-scale rebellion against digital fatigue. The Analog Rebellion isn't about rejecting technology outright; it's about reclaiming control. It's about making the conscious choice to unplug in a world that's constantly demanding our attention.

So, why now? Because we've hit a breaking point. People are waking up to the fact that digital convenience often comes at the cost of depth, meaning, and sanity. This book isn't about being anti-tech—it's about being *pro-intentionality.*

Whether you're here to dip a toe into the analog world or to go full 19th-century hermit, this book will guide you through the tools, mindset shifts, and sometimes hilarious challenges of reclaiming your attention span. Welcome to *The Great Analog Rebellion*—your mind will thank you.

ANALOG REBELLION

How I Became an Analog Revolutionary

I never planned to become an analog revolutionary. For years, I was the digital world's biggest cheerleader, helping companies migrate to the cloud and preaching the gospel of social media marketing. My blood ran in 1s and 0s—or so I thought. Then came 2018—the year my brain short-circuited from digital overload.

My days blurred into an endless scroll of emails, notifications, and algorithm-driven content. I found myself compulsively checking my phone, my attention span shrinking with every swipe. Even my dreams were interrupted by phantom alerts. It was as if my mind had been hijacked by the very technology I once championed. I couldn't remember the last time I'd held an actual pen, listened to a song without algorithmic suggestions, or looked at a photograph that wasn't filtered to oblivion. I was a cog in the machine. A very stressed-out cog.

The Analog Awakening

It started with a typewriter. A beat-up 1956 Smith-Corona I snagged from a thrift store, mostly because it matched my apartment's mid-century aesthetic. I certainly *didn't* expect it to lead me down a rabbit hole of vintage cameras, reel-to-reel tape decks, and a growing obsession with the dial-up screech of a fax machine (yes, *three* fax machines). I *definitely* didn't foresee the time I accidentally faxed my grocery list to a venture capital CEO.

But here's the thing about analog technology—it has a way of rewiring your brain. The deliberate act of loading film forces you to think twice about every shot. The inability to delete words makes you choose your words more carefully. The static between vinyl tracks becomes an unexpected moment of Zen in a world that never shuts up.

Intentional Technology Use

This book isn't a Luddite manifesto; it's not about abandoning technology but about using it more intentionally. I still have a smartphone (glued to my hand, if I'm honest), and yes, I ironically use social media to share photos of my analog adventures. *The Great Analog Rebellion* is about reclaiming balance in an increasingly imbalanced world. It's a journey into the surprising benefits of slowing down, embracing imperfection, and reconnecting with the tangible. Sometimes, the best way to move forward is to take a step backward.

ANALOG REBELLION

What You'll Discover

In these pages, you'll explore why Gen Z is trading TikTok dances for typewriter poetry (and making it go viral), why tech executives are ditching laptops for paper notebooks, and why the humble fax machine is making a comeback in the digital age. We'll dive into the art of mechanical mayhem, explore the beauty of blurry film photos, and uncover the hidden world of analog underground communities.

You'll learn how to incorporate analog tools into your daily life to boost focus and spark creativity. These simple changes can help you find a little sanity in our digitally saturated world. More importantly, you'll discover why the real analog revolution comes from a place of fun.

Join the Rebellion

So, whether you're a digital native longing for life before smartphones, a burned-out tech worker seeking refuge in the analog world, or simply someone who misses the click-clack of real keys under your fingers, welcome to the rebellion. It might not be digitized, but I promise—it will be worth the wait.

Now, if you'll excuse me, I need to go unjam my typewriter. Again.

Part 1

THE DIGITAL HANGOVER & THE ANALOG ANTIDOTE

1

Why Now?

The Rise of the Analog Underground

The Digital Hangover: How TikTok Made Typewriters Cool (And Why Millennials Are Struggling to Keep Up)

THE FIRST TIME I saw a typewriter trending on TikTok, I nearly spilled my coffee all over my mechanical keyboard collection. As a millennial who had spent years trying to escape the analog world my parents cherished, watching teenagers celebrate the very technologies I'd dismissed was like watching my past collide with the future—like a VHS tape rewinding in a Netflix era. My relationship with typewriters had always been complicated—like an old acquaintance who unexpectedly turns up and somehow makes the night more memorable. There I was, a reformed digital marketer who once preached the gospel of automation, now playing typewriter whisperer to a generation that probably thought 'carbon copy' was a clothing brand.

The Unexpected Analog Boom

Last summer, this digital-analog paradox came to life in the most unexpected way. What I'd planned as an intimate typewriter workshop at my local coffee shop turned into something that felt like a flash mob for the mechanically inclined. I walked in to find thirty-seven teenagers and young adults, many clutching vintage typewriters they'd bought at thrift stores. When I asked how they'd discovered typewriters, the answers were unanimous: TikTok. From viral typing videos to aesthetic letter-writing posts, Gen Z had turned analog tech into a digital sensation.

As I helped them navigate jammed keys and faded ribbons, I couldn't help but laugh at the irony—a former digital marketer teaching smartphone-addicted teens how to use technology that was obsolete before they were born. The air filled with the symphony of mechanical keystrokes, punctuated by occasional gasps of delight when someone successfully changed a ribbon or mastered the art of the carriage return. These digital natives were discovering something that no touchscreen could replicate: the satisfying resistance of keys that pushed back, the permanence of words struck onto paper, and the beautiful finality of thoughts that couldn't be deleted with a casual backspace.

Science Behind the Analog Appeal

Unlike the frictionless, easily editable nature of digital writing—where revisions are instantaneous and impermanent—analog tools like typewriters demand a deliberate approach. Each keystroke becomes a tangible mark, and errors are not effortlessly erased but instead integrated into the creative process. This dynamic mirrors findings from a seminal 2014 study by Mueller and Oppenheimer, published in *Psychological Science*[1]. Their research revealed that students who took notes by hand, rather than typing on laptops, retained information more effectively and engaged in deeper cognitive processing. By forcing the brain to slow down, summarize ideas, and prioritize key concepts (instead of passively transcribing words verbatim), handwriting strengthened memory connections and conceptual understanding. While the study focused on handwriting, the broader principle—that physical, less-editable mediums foster intentionality and cognitive engagement—resonates with the resurgence of typewriter culture among TikTok users, who are drawn to the tool's unapologetic permanence and focus.

A Revolution, Not a Trend

By the end of the workshop, I realized the analog revival wasn't just a trend; it was a revolution—one scaling faster than a viral dance challenge, proving that analog tools weren't just surviving but thriving in a digital-first world. These kids weren't rejecting technology—they were redefining its role in their lives, finding ways to make the old and new coexist in a wonderfully chaotic harmony. They were teaching me that maybe the future isn't about choosing between analog and digital, but about understanding when each serves us best.

This resurgence isn't merely aesthetics or nostalgia; but something deeper - a way to create authenticity while seeking tangible experiences that are limitations are viewed as creative freedom.

---

1. Mueller, P. A., & Oppenheimer, D. M. (2014). The pen is mightier than the keyboard: Advantages of longhand over laptop note taking. *Psychological Science, 25*(6), 1159–1168.
https://doi.org/10.1177/0956797614524581

ⒶⓃⒶⓁⓄⒼ ⓇⒺⒷⒺⓁⓁⒾⓄⓃ

The TikTok Effect: How Social Media Revolutionized Analog Appeal

The story of social media platforms is an ironic one, where users on TikTok, are sharing vintage cameras while showing someone loading the device. Complete with the winding sound anyone over 35 years old can recall, the video had millions of views, while teenagers begged their parents to dig out the camera.

These "analog activities" have seen themselves be transformed into small snippets of digital content. But is digital truly transforming it, if users share their first blank photos taken on their digital devices, and then discuss how they want to do it in real life?

TikTok created digital communities for people to share, but it might've unintentionally given way to a new life in the analog space.

ⒶⓃⒶⓁⓄⒼ ⓇⒺⒷⒺⓁⓁⒾⓄⓃ

Generational Divide: Why Gen Z is Embracing What Millennials Rejected

Here's the thing that keeps me up at night (besides the constant whirring of my vintage fan collection): we are living through the greatest technological plot twist since someone decided to put a clock in a microwave. Millennials spent their youth desperately trying to escape the analog world, only to watch Gen Z enthusiastically resurrect the very technologies we abandoned. It's like throwing out a pair of jeans, only to have a teenager spend $150 to buy the same pair—but now with 'authentic' holes in them.

Last month, I saw this generational divide play out at a local flea market, where a group of millennials eyed vintage tech with nostalgic hesitation while Gen Z buyers snapped up typewriters like they were the latest gaming consoles. One side approached these relics with the same cautious respect you'd give a landmine, while the other fearlessly dismantled typewriters and gushed over the 'aesthetic' of correction fluid—yes, the same correction fluid we used to sniff in high school (not our proudest moment).

When I carefully explained how to avoid paper jams, a 16-year-old participant interrupted to show me how she'd already modified hers with custom keys for emoji symbols. When I mentioned the hassle of having to retype entire pages due to mistakes, she gave me a look like I'd just discovered fire and said, 'That's the whole point—it makes you think before you speak. Unlike Twitter.'

For millennials, these machines are seen as the constraints we fought to escape. They're tools for mindfulness, authenticity, and for Gen Z, rebellion against the algorithm-driven, infinitely editable digital world.

Take my recent experience at a local record store. I watched in amazement as a group of teenagers spent hours flipping through vinyl albums, debating the merits of different pressings. When I mentioned that they could find all this music on Spotify, they looked at me like I'd suggested replacing their pets with robot dogs. 'But then you don't get to experience the whole thing,' one explained, cradling a used album that probably cost more than a year's worth of streaming. 'Plus, you can't post a screenshot of a playlist on your wall.'

It's not just about aesthetics. Gen Z's embrace of analog technology reveals a deeper yearning for something many millennials are only now beginning to understand: the value of friction. While we spent years trying to eliminate every inconvenience from our lives, they're actively seeking out technologies that force them to slow down, make deliberate choices, and accept imperfection.

The ultimate irony? Millennials are now learning about analog technology from the very generation we thought we'd be teaching about the digital world. Last week, a Gen Z colleague had to show me how to properly clean a vinyl record—a skill I somehow missed while being too busy celebrating its supposed obsolescence. As I watched her methodically demonstrate the process, I couldn't help but wonder if maybe we were too quick to dismiss these 'outdated' technologies in our rush toward digital utopia.

Maybe that's the real lesson here—progress isn't always a straight line. Sometimes, the best way forward is to take a thoughtful step backward. Or, as my teenage workshop participant so eloquently put it: 'Sometimes you need to slow down to speed up, you know?' Yes, I do know—I just needed a generation born into the digital age to remind me.

ANALOG REBELLION

The Authenticity Factor: Trading Digital Filters for Mechanical Imperfections

Let's talk about authenticity—that elusive quality we're all supposedly seeking while applying the 'Rise and Grind' filter to our morning coffee photos. But you know what's truly authentic? The way a typewriter's 'E' key sticks slightly to the left, creating a distinctive pattern as unique as a fingerprint. Try replicating that with an Instagram preset.

I discovered my obsession with mechanical imperfection during what I now call *The Great Darkroom Disaster of 2022.* There I was, attempting to develop my first roll of film in a makeshift darkroom (a.k.a . my bathroom with towels stuffed under the door), when I accidentally exposed half the roll to light. I was devastated—until I saw the results. Ghostly images with light leaks created an ethereal quality that no digital filter could match. My mistakes had become my signature.

The Beauty of Imperfection

This got me thinking: why are we so drawn to these mechanical quirks in an age where digital perfection is just a tap away?

Psychological research supports this craving for imperfection: Studies on wabi-sabi aesthetics and material nostalgia suggest that flaws in analog tools—like a typewriter's stuck key or the crackle of a vinyl record—trigger emotional resonance by contrasting with the sterile perfection of digital interfaces. For instance, Van Tilburg et al. (2015)[2] found that tactile, imperfect objects evoke stronger autobiographical memories, while Hagtvedt & Patrick (2011)[3] showed that minor flaws enhance perceptions of authenticity. In an age of algorithmically curated 'perfection,' these imperfections become acts of rebellion—a way to reclaim humanity from the machine.

It's why we love cracked pottery, handwritten notes, and, apparently, why a typewriter's broken 'E' key can feel like a badge of honor.

We're bombarded with flawless, algorithm-approved aesthetics, making a smudged typewriter key or an out-of-focus Polaroid feel like a breath of fresh air. We're not just tolerating imperfection—we're actively seeking it out.

Technology That Talks Back

Last month, I watched a group of design students have an existential crisis over a jammed typewriter ribbon. Instead of getting frustrated, they spent hours documenting how each stuck key created unique ink patterns on the paper. *"It's like the typewriter is collaborating with us,"* one student mused while carefully preserving what most would consider a ruined page. They weren't wrong—there's something compelling about technology that pushes back, that refuses to be perfectly predictable.

One of the most well-received images I've come across wasn't a carefully curated desk setup or a color-coordinated bookshelf—it was a blurry Polaroid of a cat attacking a roll of film. The photo itself was objectively terrible, with weird color shifts and a thumb

---

2. Van Tilburg, W. A. P., et al. (2015). Nostalgia's place among self-relevant emotions. *Cognition and Emotion, 29*(4), 702-719.

3. Hagtvedt, H., & Patrick, V. M. (2011). Turning art into mere reality: The effect of imperfect reproduction on judgment of art. *Empirical Studies of the Arts, 29*(2), 199-218.

partially visible in the corner. Yet something about its unpolished authenticity resonated in a way no perfectly composed digital shot ever did.

The Art of Letting Go

But here's what really gets me: these mechanical imperfections aren't just aesthetic choices—they teach us something valuable about embracing uncertainty and impermanence. When you can't just hit 'undo' or apply a filter, you start seeing mistakes differently. That coffee stain on a typewritten letter? It's not a flaw—it's proof that the object exists in the real world, with a history and a story to tell.

I recently started hosting what I call *Imperfection Workshops*, where people bring their digital photos to recreate them using analog methods. The transformation is fascinating—watching someone go from obsessing over getting the perfect digital shot to deliberately embracing light leaks and double exposures. One participant, a social media manager by day, broke down in tears after developing her first roll of film. *"These photos look nothing like I planned,"* she said, *"and they're better for it."*

The beauty of mechanical imperfections is that they aren't glitches—they're signatures. Every jammed typewriter key, every scratched vinyl record, every light-leaked Polaroid isn't just a flaw—it's proof that something real happened. No two typewriters age exactly the same way. Every vintage camera has its own quirks. Even identical record players develop distinct personalities over time. It's as if each piece of analog technology slowly becomes more itself with every use and misuse.

Permission to Be Imperfect

And maybe that's what we're really trading digital filters for—not just mechanical imperfections, but the permission to be imperfect ourselves. Just like those design students who turned a jammed typewriter into an art experiment, or the social media manager who found unexpected beauty in her flawed film photos, we're learning that the best moments often come from the mistakes we can't edit away.

So maybe that coffee stain on a typewritten letter isn't a mistake—it's a reminder that some things in life don't need a backspace button. And that's exactly why they matter. In a world where we curate and filter every aspect of our digital presence, there's something profoundly liberating about embracing technologies that remind us that perfection isn't always the point. Sometimes, the best stories come from the mistakes, the accidents, and yes—even the coffee stains.

ⒶⓃⒶⓁⓄⒼ ⓇⒺⒷⒺⓁⓁⒾⓄⓃ

The Psychology of Permanence: Why Mistakes Make Better Writers

There's something deliciously terrifying about typing on a machine that doesn't have a delete key. It's a curious mix of creativity and anxiety—on one hand, the permanence forces you to think more carefully before committing words to paper; on the other, the fear of an unfixable mistake looms large, adding an exhilarating tension to the process. It's like walking a tightrope without a net, except instead of plummeting to your death, you risk having to start your entire manifesto over because you typed 'revoltuion' instead of 'revolution.' And let me tell you, nothing teaches you the value of proofreading quite like having to retype an entire page because you got too cocky with your spelling.

The Great Novel Attempt of 2021

I learned this lesson the hard way during what I now refer to as *The Great Novel Attempt of 2021.* It was a trial by fire that made me rethink the way permanence influences our creative process. What started as a frustrating exercise in futility became a revelation about how constraints force us to be more intentional with our words and decisions. Armed with my trusty 1956 Royal Quiet De Luxe and enough coffee to fuel a small army, I decided to write my first chapter without any drafts or outlines. Four hours and seventeen sheets of crumpled paper later, I had produced exactly one usable paragraph and developed an entirely new appreciation for the psychology of permanent decisions.

Writing Without a Safety Net

Writing without a safety net forces you to slow down and be more deliberate. The permanence of each keystroke demands a deeper engagement with language, transforming the writing process into an exercise in mindfulness and precision.

Last week, I hosted a writing workshop where I forced a group of bloggers to trade their laptops for typewriters. The first hour was pure chaos—frustration, groans, and plenty of yanked-out pages. But as time went on, something shifted. The initial resistance gave way to focus, the chaos turned into quiet concentration, and by the end, many of them admitted they had never felt so engaged in their own writing process before.

'I actually have to plan my sentences now,' one participant said, staring at his typewriter with a mixture of respect and fear. 'It's like the words have weight again.' Another noticed she was writing shorter, more precise sentences—no more rambling paragraphs that could be edited later. The permanence of the medium had turned them all into more deliberate writers.

The Shared Space of Writing

The psychology behind this transformation is fascinating. When we write digitally, we're essentially working with infinite drafts simultaneously. Every word is provisional, every sentence temporary. It's liberating, sure, but it can also be paralyzing. How many times have you found yourself endlessly tinkering with a paragraph, never quite committing to a final version because you always have the option to change it?

Typewriters, on the other hand, force us into a state of mindful writing. It's much like the *Shared Space* concept in some European towns, such as Drachten in the Netherlands, where urban planner Hans Monderman removed traffic signs and lights to encourage more intuitive and cautious driving. The experiment led to fewer accidents as drivers, cyclists, and pedestrians became more aware of their surroundings and relied on eye contact and mutual understanding rather than rigid rules[4] . This mindful approach extends

4. Monderman, H., & Dijkstra, S. (2005). Shared Space: Rethinking the Human Environment's Street Design. *Proceedings of the Institution of Civil Engineers - Municipal Engineer, 158*(4), 239–246.

beyond writing; it cultivates patience and enhances our ability to make thoughtful choices in all areas of life.

This shift in mindset can extend beyond just the act of writing—it cultivates patience, decision-making skills, and a greater appreciation for the weight of our words. By fostering a habit of careful thought before action, it has the potential to influence how we approach problem-solving, communication, and creativity in other areas of life as well.

Mistakes Are Awesome!

But here's what really gets me: the mistakes. Oh, the beautiful, character-building mistakes. In our digital world, errors are things to be eliminated, scrubbed away without a trace. But on a typewriter, mistakes become part of the story. That X'd-out word, that visible correction, that slightly crooked line—they're all evidence of the human behind the machine, proof that this piece of writing emerged from a real process of thought and effort.

I keep a collection of my worst typing disasters pinned to my office wall—a gallery of glorious failures that remind me how each mistake taught me something. There's the cover letter where I managed to misspell my own name (twice), the poem where I accidentally created a new word by jamming two keys together (which, honestly, worked better than my original idea), and my personal favorite: the love letter where a stuck ribbon turned my heartfelt declarations into an abstract ink painting.

These imperfect pages tell a story that goes beyond their actual content. They're artifacts of effort, tangible proof of the messy, beautiful process of putting thoughts into words. And isn't that what writing should be? Not a sterile exercise in perfection, but a human endeavor, complete with all our glorious flaws and corrections?

ANALOG REBELLION

A New Perspective on Writing and Life

So maybe the next time you sit down to write something important, consider trading your delete key for a correction ribbon. Yes, you'll make mistakes. Yes, it will be messy. But those mistakes might just make you a better writer—or at least a more thoughtful one. After all, in a world where everything is constantly refined and optimized, committing to our words—flaws and all—feels like a radical act of authenticity. Perhaps the real lesson is that imperfection isn't just a necessary part of writing; it's a fundamental part of creativity and personal growth.

The Revolution of Analog

As I sit here wrapping up this chapter, my typewriter's ribbon running low and my fingers slightly stained with ink, I can't help but marvel at the beautiful absurdity of our current moment. Who would have thought that TikTok, the ultimate symbol of our shortened attention spans, would become the unlikely savior of analog technology? It's like watching vegetarians promote bacon—deliciously ironic and somehow perfectly fitting for our contradictory times.

As we venture deeper into this analog rebellion, remember that it's not about choosing sides in some grand technological war. It's about finding balance, embracing imperfection, and maybe, just maybe, learning to laugh at ourselves along the way.

So whether you're a digital native discovering the joy of vinyl records, a millennial rediscovering the typewriter you once scorned, or just someone looking for a little more authenticity in an increasingly virtual world, welcome to the revolution. The beauty of this moment isn't just in the nostalgia—it's in the way old and new are converging, creating a hybrid world where both have value.

Now, if you'll excuse me, I need to go help a group of teenagers figure out why their typewriter keeps making that weird clicking sound. Spoiler alert: they're typing on the margin stop. But hey, at least they're not asking me to explain what a fax machine is... yet. That's what Chapter 4 is for.

ANALOG REBELLION

Signs You're a Victim of the Digital Hangover

1. **You open your phone for "one quick thing" and wake up two hours later.**
    - You went in to check the weather. Now you somehow know what your middle school crush's cousin had for lunch in 2017.
2. **Your screen time report personally insults you.**
    - "You spent 9 hours and 42 minutes on your phone today. Are you okay?"
3. **You have 147 tabs open—and insist you'll get to them.**
    - Your browser is one wrong click away from a meltdown.
4. **You get anxious when Wi-Fi is slow, but won't answer texts for days.**
    - "Why is my video buffering?! Also, sorry for the delayed reply, just saw this from last Tuesday."
5. **You've forgotten how to wait in line without staring at your phone.**
    - Someone makes eye contact with you? Immediate panic.
6. **Your brain struggles to read anything longer than a tweet.**
    - A book? Sounds exhausting. Let me just skim the Wikipedia summary.
7. **You pause Netflix to check your phone—then forget what you were watching.**
    - You just double-screened yourself into complete amnesia.
8. **Your Notes app is where ideas go to die.**
    - It's a graveyard of forgotten passwords, half-baked resolutions, and cryptic reminders like "remember the thing at 2."

9. **You've rewatched The Office 12 times because picking something new is overwhelming.**
    - The paradox of endless options: too many choices, so you just give up and rewatch comfort TV.
10. **Your dreams now include pop-up notifications.**
    - Nothing like getting a phantom "Software Update Available" while trying to fly in your dream.
11. **You scroll on your phone while watching TikToks... on your phone.**
    - Because one screen at a time is no longer enough.
12. **You've mastered the art of watching TV with subtitles because you're always half-distracted.**
    - You hear dialogue but only process it if you *see* the words.
13. **You instinctively try to tap physical books to turn the page.**
    - And you're genuinely confused when it doesn't work.
14. **You experience "phantom vibrations" from your phone.**
    - Even when it's not in your pocket.
15. **You've accidentally liked a photo from 2012 while deep-stalking someone.**
    - Now you must disappear from society forever.
16. **You consider sending a text "talking."**
    - "I already texted them, why do I need to *call*?"
17. **You've tried to zoom in on a printed photo.**
    - And briefly wondered why it didn't work.

18. **You use Google for *everything,* even things you should know.**
    - "How long do I boil eggs?" *again.*
19. **You've started using AI to write your texts.**
    - Because even thinking about replies is exhausting.
20. **You refresh your inbox compulsively, even when you're not expecting an email.**
    - And you still act surprised when nothing new is there.

2

Vinyl Victory

When Spotify Crashes and Records Keep Spinning

Vinyl Victory: When Spotify Crashes and Records Keep Spinning

There's something ironic about paying a monthly subscription for access to millions of songs, only to find yourself staring at an error message during a dinner party. That's exactly what happened when a streaming failure unexpectedly led to a vinyl triumph, forever changing how I think about music ownership. What I didn't mention about that error message was that it arrived just as my friend Derek was giving an impassioned speech about how streaming services had 'democratized music.' The irony wasn't lost on anyone—except maybe Derek, who was too busy frantically refreshing his phone to notice the collective eye-rolling around the table.

But let me back up a bit. I had spent the better part of a decade building the perfect digital music library—meticulously organized playlists, carefully curated recommendations, and enough cloud storage to make a data center blush. Yet here I was, about to learn that the most reliable backup isn't always a digital safety net—it's the tried-and-true technology that never needed one.

In the great cosmic joke of modern technology, there's something satisfying about watching a $1,000 smartphone become a useless paperweight while a $5 thrift store record player saves the day. It's the kind of poetic justice that makes you question everything you thought you knew about progress.

The Dinner Party That Changed Everything

Last winter, our town experienced the biggest snowstorm in decades. I had planned a sophisticated dinner party, complete with a carefully curated Spotify playlist of jazz classics. As the power flickered and my Wi-Fi connection became increasingly unstable, my streaming music kept cutting out at the worst possible moments. In desperation, I dug out my grandfather's old record player and a dusty crate of vinyl from the basement. What started as a backup plan turned into the highlight of the evening.

My guests gathered around the turntable, taking turns selecting albums and watching the needle drop with childlike fascination. The warm crackle of vinyl filled the room as we flipped through worn album covers, sharing stories and discovering music we'd never bothered to click on in our streaming apps. The power eventually stabilized, but no one asked to switch back to Spotify. Instead, we spent the evening manually flipping records, appreciating the intentional act of choosing each album, and rediscovering the joy of music as a physical, shared experience. That night proved that the best backup is often the one that was never dependent on technology in the first place.

ANALOG REBELLION

The Unplugged Advantage of Vinyl

The thing about vinyl records is that they don't care about your Wi-Fi password. They won't suddenly log you out, demand two-factor authentication, or disappear behind a paywall overnight. They just play—reliably and without conditions. They don't need software updates, they don't get subscription fatigue, and they certainly don't send you passive-aggressive notifications about their terms of service changing. They just spin, delivering music with all the warmth and character that compression algorithms tend to strip away.

The Lost Art of Listening

And here's the kicker—while we've been chasing the promise of infinite choice through our streaming services, we've lost the art of actually listening to music. You know, the kind of listening where you sit down, read the album notes, and experience a record from start to finish, the way the artist intended. The kind of listening that doesn't involve algorithmically suggested playlists or the constant temptation to skip to the next track after 30 seconds.

Vinyl isn't just about nostalgia; it's about reclaiming the experience of music, one carefully chosen album at a time.

ANALOG REBELLION

The Analog Insurance Policy: Why Physical Media Still Matters

Let's talk about a dirty little secret of the digital age: everything you 'own' in the cloud is essentially a long-term rental, with terms and conditions stricter than a medieval landlord's contract. Remember when you 'bought' that digital movie collection? Plot twist—you actually purchased a limited license to stream those pixels until the service changes its policies, goes bankrupt, or simply forgets you exist.

I learned this the hard way when my entire digital music collection vanished during a platform migration, thanks to a botched update that wiped out user libraries with no recovery option. Poof—gone faster than my New Year's resolution to stop collecting typewriters. Meanwhile, my dad's vinyl collection from the '70s is still going strong, outliving eight different iPods and three streaming service subscriptions.

Here's the thing about physical media—it doesn't need permission to exist. Your vinyl records don't care if the internet is down. Your books don't require software updates to continue being books. And your Polaroid photos won't disappear because someone in Silicon Valley decided to 'optimize the user experience.'

ANALOG REBELLION

Why Physical Media Still Wins

- No subscription required
- No internet connection needed
- Can't be remotely deleted or altered
- Maintains value (sometimes increases)
- Can be repaired rather than replaced

The beauty of physical media isn't just its independence from digital infrastructure—it's its persistence. A scratched DVD might skip during a crucial scene, but at least it won't tell you it's 'currently unavailable in your region.' Just ask fans of classic shows and movies that have mysteriously disappeared from streaming platforms overnight due to expired licensing deals. A vinyl record might have some pops and crackles, but it won't suddenly remove your favorite track because of a licensing dispute.

I recently started keeping a 'physical media diary' to track how often my analog backups saved the day. In just one month, I counted six streaming service outages, three 'temporarily unavailable' albums, and one particularly frustrating moment when an entire season of a show I was watching vanished mid-episode. Each time, my physical media collection stepped in like a reliable old friend—one that doesn't need a Wi-Fi password to come over for dinner.

Legacy and Longevity

Perhaps the most compelling argument for physical media isn't just about reliability—it's about legacy. When my grandfather passed away, he left behind a treasure trove of vinyl records, each one telling a story through its worn sleeve and handwritten notes. Try leaving your Spotify playlists to your grandkids and see how that works out—meanwhile, my grandfather's vinyl collection still holds the stories of his youth, passed down like a tangible inheritance.

The irony isn't lost on me that as storage technology becomes more 'advanced,' it somehow becomes less reliable for long-term preservation. I can still play a vinyl record from the 1950s, but good luck accessing that WordPerfect file from 1995. Physical media isn't just an insurance policy against technical failures—it's an investment in your ability to access and enjoy the content you've collected for decades to come.

And let's talk about format obsolescence. While digital formats come and go faster than tech startup CEOs, physical media tends to stick around. Sure, you might need to hunt down a VCR for those old tapes, but at least the tapes still exist. Try finding a way to play your extensive collection of RealPlayer files from 2002.

At its core, this isn't about rejecting digital convenience—it's about recognizing that 'access' isn't the same as 'ownership,' a distinction we often overlook until it's too late. Understanding this means realizing that sometimes the best backup system is the one that doesn't need to be backed up at all. While everyone else is busy upgrading their cloud storage and juggling streaming subscriptions, I'll be here, building my physical media library one piece at a time—secure in the knowledge that my entertainment doesn't depend on someone else's servers staying online.

ⒶⓃⒶⓁⓄⒼ ⓇⒺⒷⒺⓁⓁⒾⓄⓃ

The Ritual of Vinyl: From Sleeve to Needle Drop

There's something almost sacred about the vinyl ritual, a choreographed dance between human and machine that digital streaming can't quite replicate. Streaming may be effortless, but it lacks the physical connection—the act of choosing, handling, and engaging with the music itself. A tap on a screen delivers sound, but it doesn't invite you into the process the way vinyl does. Have you ever seen someone reverently unbox a Spotify playlist? Didn't think so.

The process begins with selecting an album—not by mindlessly scrolling through a playlist, but by physically flipping through your collection, feeling the weight of each record, and admiring the artwork that some poor graphic designer probably spent months perfecting, only to have it shrunk down to a thumbnail on your phone. It's like choosing a wine at a fancy restaurant, except instead of pretending to understand tasting notes, you're pretending to know what 180-gram vinyl means—which, for the uninitiated, refers to a heavier, high-quality pressing that often results in better durability and sound fidelity.

Removing the sleeve is an art form in itself. Too aggressive, and you'll tear the paper inner sleeve—a rookie mistake I've made more times than I care to admit. Too gentle, and you'll spend five minutes doing what I call the 'vinyl shuffle,' where the record refuses to slide out, leaving you looking like you're coaxing a stubborn cat from under a couch.

Then comes the moment of truth: the inspection. Holding the record at just the right angle, where the light catches every microscopic detail, you suddenly develop the eyesight of an eagle and the attention to detail of a forensic investigator. That speck of dust? Unacceptable. That fingerprint? A personal affront to audio fidelity itself.

The Pre-Play Checklist:

- Clean the record with a carbon fiber brush
- Check the stylus for debris
- Set the correct tracking force
- Align the tonearm properly
- Hold breath (optional but recommended)

The actual needle drop is where the magic happens. It's like performing microsurgery, except instead of saving lives, you're trying not to scratch your copy of *Dark Side of the Moon* for the hundredth time. That moment of anticipation as the needle hovers over the lead-in groove is better than any cliffhanger Netflix has ever produced—think *Squid Game*'s shocking eliminations or the jaw-dropping twists in *The White Lotus.*

And then there's that sound—that beautiful, imperfect crackle before the music starts. It's the vinyl equivalent of clearing your throat before giving a speech, a reminder that what you're about to hear isn't just music; it's a performance. Each pop and click is a timestamp, marking every time this record has been played, every party it's soundtracked, and every heartbreak it's consoled.

I recently hosted a 'vinyl initiation ceremony' for some friends who had never experienced the ritual before. One of them, a dedicated Spotify user, spent ten minutes trying to figure out how to 'skip to the good part' before realizing that was entirely missing the point. By the end of the evening, she had bought her first turntable—not because the sound quality was demonstrably better (though don't get me started on that debate), but because she had fallen in love with the ritual itself.

Slowing Down and Being Present

The beauty of the vinyl ritual isn't just in its complexity—it's in how it forces us to slow down and be present. You can't just hit play and walk away. Well, you can, but then you'll have to flip the record, and if you're not there to do it, you'll end up with the world's most expensive silence. It's analog's way of saying, *Hey, maybe sit down and actually listen for a while.*

Each step of the ritual adds to the experience, creating what psychologists might call *multi-sensory engagement* but what I call *actually paying attention to your music for once.* In contrast, digital listening often turns music into mere background noise—an endless shuffle of algorithm-driven tracks that require no real interaction. The physical interaction, the visual appreciation of the album art, the satisfying click as the needle finds its groove—it all combines to create something more meaningful than just pressing play on your phone.

And yes, sometimes the ritual goes wrong. The anti-skate weight might take a vacation, or your cat might decide to use your tonearm as a scratching post. But that's part of the charm. In a world where everything is one-click convenient, there's something refreshingly human about occasionally messing up the simple act of playing a record.

The vinyl ritual isn't just about playing music—it's about creating a moment, a deliberate pause in an otherwise hyper-digital world. It's a reminder that in an age of instant gratification, some experiences are richer when they require effort and attention. It transforms listening from a passive background activity into an intentional experience. It's about remembering that sometimes the best things in life require a little effort, a bit of patience, and the willingness to accept that not everything needs to be instant or perfect.

ANALOG REBELLION

Sound Quality Showdown: Digital Compression vs. Analog Warmth

Let's talk about the elephant in the room—or should I say, the compressed elephant squeezed through a digital straw? The debate between digital and analog sound quality has sparked more heated arguments than those between streaming advocates and vinyl purists over whether convenience trumps authenticity.

I used to be that guy who'd bore everyone at parties with technical jargon about bit rates and sampling frequencies. That is, until my friend Tom shut me up with a simple demonstration. He played the same jazz recording twice—once through a high-end streaming service and once on vinyl. 'Stop thinking,' he said, 'and just listen.' The difference wasn't just in the sound; it was in the experience. The vinyl had a presence, a warmth that made the brass sound like it was being played in the room. The digital version was perfect—too perfect, like a wax fruit that looks flawless but leaves you unsatisfied.

Now, before the audiophiles sharpen their styluses, let's get something straight: digital music isn't bad. It's incredibly convenient, and modern compression algorithms are engineering marvels. But comparing digital to analog is like comparing a photograph of a campfire to actually sitting beside one—they're both valid experiences, but one is undeniably more immersive.

Digital vs. Analog: What's Really Different?

- **Digital compression removes 'unnecessary' frequencies**
- **Analog captures the full wave of sound**
- **Vinyl's imperfections add character**
- **Digital offers precise reproduction (for better or worse)**
- **Each format has its own distinct 'flavor'**

The real magic of analog sound isn't in its technical specifications—it's in what compression algorithms consider *unnecessary*. The slight buzz of a tube amp warming up, the subtle echo of sound reflecting off studio walls, the gentle variations in tape speed that give recordings their organic feel. Digital compression filters out these nuances like a minimalist interior designer tossing your grandmother's handmade quilts—erasing the warmth, depth, and slight imperfections that give music its soul and make it feel alive.

I recently set up a listening station at a local coffee shop with both vinyl and digital playback of the same albums. The reactions were fascinating. People who claimed they didn't care about audio quality found themselves drawn to the vinyl setup, often without realizing why. One person put it best: 'The digital version sounds like looking at the band through a window. The vinyl sounds like they're in the room with us.'

The Cost of Perfection

The irony of modern digital music is that we've achieved 'perfect' sound reproduction at the expense of something intangible yet essential. It's like the difference between a hothouse tomato and one grown in your backyard. The hothouse tomato is perfectly round, vibrantly red, and completely forgettable. The homegrown one might be misshapen and have a few blemishes, but it actually tastes like food.

And this isn't just nostalgia talking. The resurgence in vinyl sales isn't being driven by aging audiophiles longing for the past—it's led by younger listeners who crave a deeper, more tactile connection to music. The act of physically selecting an album, placing it on a turntable, and carefully dropping the needle offers a level of engagement that digital music simply can't replicate. In an era of endless streaming options, vinyl provides a sense of intentionality and presence that resonates with those looking to slow down and truly experience their music. They're discovering what their grandparents knew: sometimes the medium is part of the message.

ANALOG REBELLION

Vinyl's Creative Comeback

Here's where it gets interesting: many modern artists are now recording with vinyl in mind, embracing the format's 'limitations' as creative tools. They've realized that working within constraints often leads to more innovation than having infinite digital options. It's the same reason why writing a letter forces you to think more carefully about your words than firing off a quick text message.

The truth is, both formats have their place. Digital is perfect for discovering new music, curating playlists, and carrying an entire library in your pocket. But analog offers something different—not just a way of hearing music, but a way of *experiencing* it. It's the difference between scanning a menu and savoring a meal.

Choosing the Experience

So maybe the real question isn't which format sounds better, but which experience pulls you deeper into the music. Do you want background noise, or do you want to feel like you're in the room with the artist? Sometimes you want the convenience of digital, and sometimes you want to drop the needle on a record and let the warmth wash over you, imperfections and all. After all, isn't that what makes us human—our beautiful, analog imperfections?

The Ritual of Listening: How Vinyl Forces Us to Slow Down

In an age where we can summon any song with a tap, there's something wonderfully defiant about vinyl's insistence that we wait. It's a deliberate act of patience in a world obsessed with speed.

I learned this firsthand during my ill-fated attempt at a full analog lifestyle in 2019. Determined to ditch digital, I swapped my streaming services for a turntable and my father's dusty record collection. On my first evening, I nearly lost my mind when I realized I couldn't instantly skip to my favorite track. Hovering over the turntable like a caffeine-addled hummingbird, needle in hand, I tried to pinpoint the exact groove where the good part started. Spoiler alert: I scratched the record so badly it now sounds like it features backup vocals by an asthmatic cat.

But here's the thing about vinyl—it forces you to embrace the art of patience. You can't multitask while playing a record unless you consider 'getting up to flip the album' a task worthy of your LinkedIn profile. The format demands your attention, your presence, and, most importantly, your time. It's like having a high-maintenance friend who happens to make incredible music.

Why Vinyl Slows You Down

- **Records require physical interaction every 20-25 minutes**
- **Each side tells a curated story through song order**
- **Album art becomes part of the listening experience**
- **The ritual creates a dedicated listening space**
- **Imperfections become part of the story**

I recently hosted a *Slow Listening Party*, where the only rule was to sit through an entire album side without touching a phone. The results were fascinating. People actually talked between songs, discussed the music, and shared stories. One guest admitted it was the first time in years she'd listened to an album from start to finish without checking her email. Another said it reminded him of being a teenager, when music wasn't just background noise for scrolling through social media.

The beauty of vinyl's forced slowdown isn't just in the listening—it's in how it changes our relationship with time itself. When you know you'll need to flip the record in 20 minutes, you become more aware of each moment. You start to appreciate the artistry of album sequencing, understanding why certain songs flow into each other and why some need room to breathe. It's like reading a book instead of skimming headlines—sure, you could get the gist faster, but you'd miss all the good stuff in between.

And let's talk about those gaps between songs—those precious seconds of anticipation as the needle moves to the next track. In our digital world, we've nearly eliminated these moments, creating a constant stream of sound that never allows for reflection. But vinyl preserves these pauses, these musical commas that give us time to process what we've just heard before diving into the next experience.

The Power of Intentional Listening

The ritual of vinyl listening isn't just about slowing down—it's about creating intention. When you physically place a record on a turntable, clean it, and carefully lower the needle, you're making a conscious choice about what you want to hear. It's the difference between mindlessly grabbing fast food and cooking a meal from scratch. Sure, both will feed you, but only one makes you appreciate every bite.

I've started keeping what I call a *Vinyl Journal*, documenting how each album feels different when listened to in its entirety. Songs I used to skip have become favorites, hidden details emerge from familiar tracks, and B-sides reveal themselves as unexpected gems. It's like finally paying attention to the quiet friend at the party and discovering they're actually the most interesting person in the room.

The slow listening movement isn't about rejecting digital convenience—it's about recognizing that sometimes the best way to experience music is to let it unfold at its own pace. Streaming encourages skipping, shuffling, and passive background noise, but vinyl demands presence, transforming listening into an active and immersive experience. In a world that's constantly trying to speed things up, vinyl stands as a reminder that some experiences can't—and shouldn't—be rushed. It's analog's way of saying, *Sit down, shut up, and listen. The cat videos can wait.*

ANALOG REBELLION

In this chapter, we've explored how the vinyl renaissance isn't just about nostalgia—it's about rediscovering what we lost in our mad dash toward digital convenience. From power outages that turned into impromptu listening parties to the therapeutic value of those static-filled pauses between tracks, we've seen how analog audio creates experiences that no streaming service can replicate.

What started as my personal journey into vinyl collecting has evolved into something far more meaningful: a reminder that sometimes the best technology isn't the newest or the most efficient, but the one that makes us slow down and actually experience the moment. While my digital playlists gather virtual dust in the cloud, my record collection continues to grow, each album adding not just music but stories, memories, and yes, even a few spectacular failures to my life.

The vinyl resurgence isn't about rejecting progress—it's about recognizing that not all progress moves in a straight line. Sometimes you have to spin in circles at 33⅓ RPM to find what you're looking for. As we close this chapter, I'm reminded of what my friend Tom said during that fateful listening session: *Digital is like fast food—convenient, consistent, and completely forgettable. Vinyl is like cooking a meal from scratch—it takes more effort, makes more mess, but creates memories that last longer than the food itself.*

So next time your Wi-Fi drops out during a dinner party, don't curse the digital gods—consider it a reminder that technology isn't always as reliable as we assume. Instead, take it as an opportunity to drop the needle on a record and rediscover a listening experience that doesn't depend on an internet connection or a subscription service. And if you happen to scratch your favorite album while attempting to DJ like it's 1977, just remember—those imperfections are now part of your story, like the coffee stain on your favorite book or the dent in your grandmother's old cooking pot.

Whether you're a seasoned vinyl veteran or a curious digital native dipping your toes into analog waters, remember that the joy isn't in perfect playback—it's in the ritual, the warmth, and yes, even in those moments when your cat mistakes your tonearm for a new toy. Because in the end, it's not about choosing between analog and digital—it's about finding the sweet spot where technology serves our humanity, not the other way around.

And hey, if nothing else, you've learned that a record collection isn't just a better conversation starter than a Spotify playlist—it's a testament to the moments, emotions, and stories that music can hold in its grooves, waiting to be rediscovered with every spin. Though I should probably warn you—once you start down the vinyl rabbit hole, your bank account may never forgive you. But your soul? Your soul will thank you every time that needle drops.

ⒶⓃⒶⓁⓄⒼ ⓇⒺⒷⒺⓁⓁⒾⓄⓃ

You Know You're a Vinyl Addict When...

1. **You refuse to listen to music unless it "breathes."**
   - If it doesn't crackle, it's not real.
2. **You spend more time cleaning your records than cleaning your house.**
   - Your vinyl collection has its own dedicated microfiber cloths.
3. **You judge people based on their turntable brand.**
   - "Oh, you use a Crosley? That's cute."
4. **You own more albums than pairs of socks.**
   - Priorities.
5. **You consider flipping a record a form of exercise.**
   - Who needs a gym when you have a double LP?
6. **You get physically uncomfortable when someone touches a record incorrectly.**
   - "Did you just put your fingers on the grooves?!"
7. **You've spent more on vinyl storage than actual furniture.**
   - Who needs a couch when you have Kallax shelves?
8. **You think digital music sounds "too perfect."**
   - If it doesn't have a little static, what's the point?
9. **You have a favorite record press.**

- And you can hear the difference.

10. **You've rebought the same album five times because of different pressings.**
    - And you regret nothing.
11. **You've spent hours in a record store and still felt like you needed more time.**
    - "Just one more crate..."
12. **You panic when someone stacks records vertically.**
    - "Are you trying to warp them?!"
13. **You have at least one framed album cover as wall art.**
    - Bonus points if it's autographed.
14. **You own a backup stylus—just in case.**
    - Because you'd rather be prepared than suffer the silence.
15. **You have a dedicated ritual for playing records.**
    - Clean, cue, lower the needle, close your eyes—magic.
16. **You get into heated debates over 180g vs. standard weight pressings.**
    - And you have strong opinions.
17. **You've corrected someone who called vinyl "aesthetic."**
    - "It's a lifestyle."
18. **You've lost an afternoon alphabetizing your collection.**
    - And still debated changing the system.
19. **You feel a deep sense of betrayal when an album skips.**

    - "Not you, Dark Side of the Moon!"

20. **You don't trust people who don't own at least one vinyl record.**
    - "How do you even listen to music?"

ANALOG REBELLION

Final Thoughts

If you nodded along to most of these, congratulations—you are officially part of the vinyl addiction club. Don't worry, you're in good company. Now, go flip that record before it sits too long on the platter!

3

The Analog Creative Renaissance

Writing, Journaling, and Drawing in a World of Distractions

Writing, Journaling, and Drawing in a World of Distractions

THE MODERN CREATIVE PROCESS has been hijacked. It didn't happen all at once—it was a slow, insidious takeover, like when you realize your five-minute social media break has turned into a three-hour deep dive into conspiracy theories about whether birds are real. Creativity has become less about actual ideas and more about gaming the system. Between endless notifications, algorithm-driven feeds, and the constant pressure to create content, we've become less like artists and more like data analysts optimizing for clicks.

We don't create anymore—we curate, we optimize, we algorithm-proof our very thoughts like we're filing tax returns. We've gone from writing heartfelt journal entries to crafting Instagram captions that make us sound deep but not *too* deep—just enough to get engagement, but not enough to make anyone spiral into an existential crisis, quit their job, and start wearing only linen while living off the grid. Meanwhile, our attention spans have been reduced to whatever time it takes for a goldfish to give up on life, or for a TikTok user to swipe past your carefully edited post with the speed of a caffeinated squirrel.

That's where analog tools come in—not just as aesthetic choices, but as weapons in the fight for **deep, focused creativity.** They provide a way to slow down, be more deliberate, and re-engage with the physical world in ways that digital tools simply cannot replicate. Analog creation isn't just about nostalgia—it's about reclaiming attention, intention, and creative flow.

ANALOG REBELLION

The Sensory Experience of Analog Creation

Digital tools are designed to be **invisible.** You type on a laptop keyboard, and the words appear on a glowing screen, but you don't really **feel** them. The experience of analog creation is entirely different—it engages **your senses** in a way that digital tools never can.

- **The tactile feedback of pressing typewriter keys** is different from the shallow tap of a keyboard. There's a certain "thunk" that lets you know you're actually doing something.
- **The smell of ink and paper** in a notebook activates memory in a way a screen never will. It's like aromatherapy, but for writers with deadlines.
- **The physical act of sharpening a pencil** or dipping a brush in paint makes creation feel **deliberate.** Also, let's be honest—sharpening a pencil makes you feel like you're about to do something important.
- **The sound of a typewriter's carriage return** or the scratch of a fountain pen adds an auditory layer to creativity. It's an ASMR experience before ASMR was a thing.
- **The weight of a book or a journal in your hands** provides a sense of permanence that digital files lack. A laptop battery dies; a journal just sits there, being all reliable.

Analog isn't just a tool—it's a **full-body experience.** That's why so many creatives are turning back to it: **because it feels good.**

ANALOG REBELLION

The Typewriter Diaries: What Happens When You Can't Hit Delete?

I thought using a typewriter would make me feel like Hemingway. Instead, I felt like a medieval scribe being punished for my sins.

For one week, I ditched my laptop and attempted to write using only a vintage Olympia SM9. Here's what happened:

- I **typed more slowly**, but somehow, my thoughts flowed better.

- I **stopped over-editing** every sentence before moving forward. No backspace meant no neurotic rewrites.
- I **felt real commitment** to every word—because fixing mistakes was a pain. I had to own my typos like they were questionable life choices.
- I **embraced imperfection**, because backspacing wasn't an option. Turns out, writing "hte" instead of "the" won't kill you.
- I **noticed my ideas developed more organically**, rather than being polished into overprocessed perfection.
- I **became hyper-aware of word choice**—each keystroke felt intentional, like chiseling thoughts into stone.

By the end, I realized that **friction forces clarity.** The resistance of analog tools compels creators to think ahead, make deliberate choices, and embrace the process rather than obsessing over the final product. The inability to delete made me more decisive. There was no second-guessing, no infinite tinkering. Instead of perfect, my writing became **deliberate.**

Could I live like this forever? Absolutely not. But for creative breakthroughs? There's something magical about writing in a way that forces you to **commit.**

ANALOG REBELLION

The Notebook Comeback: Why Paper is Beating Productivity Apps

At some point in the last decade, we convinced ourselves that true productivity required downloading 20 different apps, each promising to revolutionize our efficiency—while quietly ensuring we'd spend more time organizing tasks than actually doing them.

Enter **the bullet journal**—a method so simple, yet so effective, that it has made people feel guilty for abandoning paper in the first place. Handwritten to-do lists and journals aren't just surviving; they're thriving. Why? Because they do something no digital tool can:

- **They force you to focus.** No notifications. No auto-correct. Just you and your thoughts.
- **They make tasks feel more tangible.** Crossing off an item with a pen is infinitely more satisfying than clicking a checkbox.
- **They don't crash.** Your notebook never suddenly refuses to open unless you update your operating system.
- **They aren't spying on you.** Ever get a targeted ad for something you wrote in your notes app? Yeah, that doesn't happen with paper.

Productivity apps may promise efficiency, but analog tools promise something better: **peace of mind.**

ANALOG REBELLION

Sketching, Doodling, and the Rebellion of Handcrafted Art

In a world where AI can now generate art faster than you can say "existential crisis," creating something by hand has become an **act of defiance.** Once upon a time, sketching was just a pastime. Now, it's a protest.

Artists are ditching digital perfection in favor of **physical mediums—charcoal, watercolor, film sketching—not because they have to, but because imperfection is human.**

- **Hand-drawn sketches feel raw, real, and deeply personal.** They tell a story beyond the subject.
- **There's no "undo" button.** You either make it work or turn that mistake into

a "creative choice."

- **Physical art can't be stolen by AI models (well, not easily).** If a robot wants to copy your work, it's going to have to figure out how to hold a pencil first.
- **The feeling of ink soaking into paper is irreplaceable.** It's messy. It's permanent. It's alive.

When creativity becomes too easy—too editable, too flawless—it loses something. The messy, frustrating, imperfect process of analog art? **That's where the magic happens.**

ⒶⓃⒶⓁⓄⒼ ⓇⒺⒷⒺⓁⓁⒾⓄⓃ

Creativity and Memory: Why Writing by Hand Makes Ideas Stick

Ever notice how you remember things better when you write them down? It's not just an old-school trick—there's actual science behind it. We touched on this back in Chapter 1, but let's dig deeper because, honestly, it deserves more attention.

Studies have shown that students who take notes by hand **retain more information and understand concepts better** than those who type. That's because writing by hand requires:

- **More cognitive engagement** (you have to actively think about what to write).
- **More muscle movement**, which reinforces memory.
- **Slower processing**, which means you filter information instead of transcribing mindlessly.
- **Physical interaction with the page**, which connects ideas more meaningfully.

This is why **authors like J.K. Rowling, Quentin Tarantino, and Neil Gaiman** still write first drafts by hand. They know that **slowing down fuels creativity.**

ANALOG REBELLION

Fun List: Signs You've Gone Too Deep into the Analog Life

1. You own more notebooks than you will ever fill in your lifetime—but you keep buying more because *this one* might finally make you organized.
2. You scoff when someone calls a typewriter "vintage"—because your prized model is older than their grandmother.
3. Your fingers are permanently ink-stained, and you consider it a mark of honor.
4. You refer to digital notes as "soulless pixels" and insist that all ideas must be written on *real* paper.
5. You have a mental catalog of fountain pen nib sizes and could give a TED Talk on the best ink for journaling.
6. You spend more time hunting for the *perfect* stationery than you do actually writing letters.
7. You've attempted (and failed) to convert at least three friends to typewriter life.
8. You get visibly angry when someone suggests using AI to "improve" their handwriting.
9. Your bookshelf has an entire section dedicated to "the lost art" of something—letter writing, calligraphy, bookbinding.
10. You've started referring to WiFi as "the digital overlord" and look longingly at candlelit desks in period films.
11. You hoard typewriter ribbons like they're a rare currency for the impend-

ing tech apocalypse.

12. You refuse to write with anything but a fountain pen, even if it means carrying a tiny bottle of ink everywhere like an eccentric wizard.

13. You have strong opinions about which typewriter font is superior, and you will die on the hill that Courier is the only acceptable choice.

14. You get irrationally upset when someone rips a page out of a notebook instead of cleanly removing it along the perforation.

15. You have seriously considered buying a wax seal for your letters—despite the fact that you send maybe one handwritten letter a year.

16. You feel a deep sense of accomplishment when a journal is completely filled, as if you've completed an ancient rite of passage.

17. You refuse to highlight books because "they should remain pristine," but you will write extensive notes in the margins with a very specific mechanical pencil.

18. You have at least one pocket notebook at all times, just in case a world-changing idea strikes while you're standing in line at the post office.

19. You've developed a slight paranoia that digital files will one day vanish in an apocalyptic server crash, but your handwritten notes will survive the ages.

20. You secretly believe that using analog tools makes you a more thoughtful, intellectual, and possibly more mysterious person—but you'd never admit it out loud.

Final Thoughts: The Slow, Messy, and Beautiful Art of Creating by Hand

Analog tools aren't about rejecting digital life—they're about **reclaiming the creative process** from distractions, perfectionism, and automation.

- When you **write by hand**, you **own** your words.
- When you **draw on paper**, every line **matters.**
- When you **use a typewriter**, you can't edit forever—you just **move forward.**
- When you **journal**, it's an intimate space that no algorithm can manipulate.
- When you **fill a notebook**, you create a tangible record of your thoughts, unfiltered and uncurated.

Maybe that's the lesson of analog creativity: **stop chasing perfection, embrace the mess, and make something real.** Just like my week with the typewriter forced me to commit to every word—typos, awkward phrasing, and all—analog creation encourages us to trust the process instead of endlessly revising. The imperfections, the rough edges, and the spontaneity—they're not just acceptable; they're the point. Because at the end of the day, the best creative work isn't the one that's been polished to oblivion—it's the one that actually gets finished.

And honestly, who doesn't love the sound of a typewriter bell dinging at the end of a perfect sentence? It's the closest thing a writer gets to applause.

4

Slowing Down to Speed Up

The Zen of Film Photography

THERE ARE FEW THINGS more humbling than thinking you're an artist and then realizing you've just spent $30 developing 24 blurry photos of your thumb.

Welcome to the world of film photography—where every click is a gamble, every shot is an exercise in patience, and every roll of film has the potential to be a masterpiece or a complete and utter waste of time. And money. Mostly money.

ANALOG REBELLION

The Slowest Way to Take the Worst Photos of Your Life

Few things in life are as entertaining as watching someone raised on iPhones try to use a film camera for the first time. It's like watching a baby deer try to walk—wobbly, confused, and doomed to fail multiple times before getting it right.

I once handed a disposable camera to a Gen Z friend, and she immediately flipped it over, looking for the screen. I had to explain that you wouldn't see the results for at least a week, which led to the same expression of existential dread people get when they realize their battery is at 2% with no charger in sight.

Another time, I gave a disposable camera to my cousin, who, after carefully composing a shot, tapped the back of it furiously before looking at me in pure horror. 'Where did the picture go?' she gasped. The concept of waiting for photos to be developed was completely foreign to her—like asking someone in 1995 to predict TikTok.

Then there was my friend Jerome, who excitedly took a roll of what he was convinced were 'incredible' street photography shots, only to realize he never actually loaded the film. He had essentially spent an entire afternoon pretending to be a photographer, like a kid making race car noises in a cardboard box.

One of my favorite stories comes from a college student who bought a vintage medium format camera and proudly told me he was going to take 'some sick Polaroids' with it. When I tried to explain that medium format film doesn't develop instantly, he looked at me like I had just told him Santa wasn't real.

This new generation of photographers approaches film cameras the same way someone approaches an old rotary phone—fascinated but confused, slightly afraid, and deeply resentful of how much harder life used to be. Watching them grapple with concepts like manual focusing and ISO settings is a masterclass in patience.

But despite their initial struggles, something magical happens: They start to love it. The frustration turns into curiosity, the hesitation into excitement. The first time they successfully wind a roll of film, they hold the camera like they just tamed a wild animal. And when their developed prints finally come back, it's like Christmas morning—except instead of getting exactly what they asked for, they get something totally unexpected, flawed, and yet somehow perfect.

The first thing you need to know about shooting on film is that it's a commitment. A deeply frustrating, expensive, and wildly unpredictable commitment, like adopting a cat with a mood disorder. Unlike digital photography, where you can rapid-fire 600 shots of your brunch and delete the ones where the avocado toast looks *slightly* less photogenic, film forces you to think. Every frame costs money, every shot requires deliberation, and—God help you—every mistake is permanent.

For those of us raised in the era of infinite cloud storage and AI-enhanced selfies, this is a level of responsibility we are simply not equipped for. You don't just take a photo. You *prepare* to take a photo. You check the light meter, consider your aperture, wonder if you need a filter, double-check the focus, and by the time you're ready to click the shutter, the moment is long gone.

And when you finally do capture something magical? You won't know it. Not for days. Maybe weeks. Maybe ever, if the lab loses your film or—*shudder*—your entire roll comes back blank because you accidentally loaded it incorrectly.

ANALOG REBELLION

Why Film Photography Has a Cult Following in the Digital Age

In an era where you can deepfake an entire person, erase blemishes with a swipe, and make your cat look like a Renaissance painting, film photography stands as a beacon of imperfection. It's the vinyl records of photography—quirky, nostalgic, and requiring just enough effort to make you feel superior to your fully digital friends.

Think about it: We live in a world where people spend money on apps to add *fake* film grain and light leaks to their digital images. That's like using a Bluetooth typewriter—it defeats the purpose entirely.

Film cameras are the photographic equivalent of Stranger Things: they remind us of a time when life was simpler, slightly out of focus, and tinged with the ever-present possibility of disaster. It's no wonder hipsters, artists, and Hollywood directors alike have clung to film despite the existence of digital cameras that can see in the dark, make toast, and probably perform minor surgery.

ANALOG REBELLION

The Film Photographer's Emotional Journey: A Step-by-Step Breakdown

Shooting on film is an emotional rollercoaster. If you've ever tried it, you know the feelings cycle through very specific stages. Here's the complete psychological breakdown:

1. **Unbridled Enthusiasm** – You just bought a film camera. You are convinced you are about to produce works of staggering genius. You picture yourself strolling through misty streets like a noir detective, capturing Pulitzer-worthy shots with every click.

2. **First-Timer Confidence** – You load the film (incorrectly, but you don't know that yet), adjust a couple of dials you barely understand, and take your first shot. You feel like a pro. You have *the eye*.

3. **Oh No, Wait—How Does This Work?** – You realize there's no auto-focus, no digital screen, no instant feedback. You start second-guessing everything. Is this

even working? What if you're just pressing buttons and making clicking noises like a child pretending to be a photographer?

4. **The "Did I Just Waste a Shot?" Anxiety** – Every time you press the shutter, a wave of regret washes over you. Was that photo worth the 75 cents it will cost to develop? Did you actually frame it right, or will this be a blurry disaster?

5. **The Overly Careful Phase** – You take FOREVER to compose shots, adjusting settings, waiting for perfect light, consulting the nearest wizard for advice. You start treating each shot like a rare and finite resource (because, unlike digital, it *is*).

6. **The "Screw It, Let's Shoot" Phase** – You get tired of overthinking and just start snapping. Maybe too much. Suddenly, you're down to your last three frames, and you have nothing but pictures of buildings you don't actually care about.

7. **The Agonizing Wait** – You drop the film off at a lab, and suddenly you understand what it was like waiting for test results in the 1800s. You check your phone compulsively, even though you *know* the results won't magically appear in your email.

8. **The Moment of Truth** – You pick up your prints or scan your negatives, hands trembling like you just cracked a safe.

    - **Best-case scenario:** Some of the shots are *actually good*. You are a genius. You should open a gallery.

    - **Worst-case scenario:** Everything is underexposed, overexposed, or inexplicably blurry. You stare at your failed prints like a tragic artist, contemplating throwing your camera into the sea.

9. **Selective Memory Kicks In** – Even if only *one* picture turned out well, you convince yourself film is amazing and totally worth it. You immediately buy more rolls and start the cycle all over again.

ⒶⓃⒶⓁⓄⒼ ⓇⒺⒷⒺⓁⓁⒾⓄⓃ

The Internet's Love Affair with Film Photography

Social media has played a weird role in reviving film photography. Instagram, TikTok, and YouTube are flooded with creators sharing their love of expired film stocks, weird Russian cameras, and DIY developing techniques that involve coffee, tea, and, for some reason, Vitamin C tablets. (Yes, people have tried to develop film in orange juice. No, it does not work.)

There are entire accounts dedicated to "Film Camera ASMR," where people just click shutters and wind film advance levers. It's oddly satisfying, like watching Bob Ross paint or listening to a campfire crackle.

Meanwhile, celebrities have embraced film photography with the same enthusiasm as overpriced oat milk lattes. Zendaya, Robert Pattinson, and even *Billie Eilish* have been spotted lugging around film cameras like it's 1994.

Before we dive into the technical breakdown of film cameras, let's acknowledge one universal truth: every film photographer has a love-hate relationship with their gear. Some cameras feel like they were engineered by gods; others seem like a cruel joke from the 1970s. Understanding the different types of film cameras can help you find the perfect balance between nostalgia and sheer frustration.

ⒶⓃⒶⓁⓄⒼ ⓇⒺⒷⒺⓁⓁⒾⓄⓃ

The Many Faces of Film Cameras: A User's Guide to Joy and Suffering

Not all film cameras are created equal. Some are beautiful pieces of engineering that make you feel like a 1970s war photographer; others exist solely to test the limits of your patience. Let's break them down:

1. **The SLR (Single-Lens Reflex)**
    - These are the workhorses of film photography—legendary, reliable, and capable of producing stunning images if you know what you're doing. And if you *don't* know what you're doing? Congratulations, you've just taken an entire roll at f/22 in a dimly lit room.
    - Pros: Full control, beautiful optics, feels professional.
    - Cons: Requires actual knowledge of photography, bulky enough to double as a self-defense weapon.
2. **The Point-and-Shoot**
    - The film camera equivalent of a frozen pizza—low effort, decent results, but nowhere near what a real photographer would consider gourmet.
    - Pros: Auto-focus, auto-exposure, easy to use.
    - Cons: Your grandmother's old one probably still has an expired roll of film inside.
3. **The Rangefinder**
    - If SLRs are the reliable sedans of film photography, rangefinders are the vintage sports cars—beautiful, finicky, and likely to make you cry when they break.
    - Pros: Compact, whisper-quiet shutter, great for street photography.
    - Cons: Requires a separate focusing mechanism, which means you will *definitely* miss a once-in-a-lifetime shot because you were too busy trying to align two tiny images inside a viewfinder.
4. **The Disposable Camera**
    - The only film camera that comes preloaded, is designed to be thrown away, and somehow still costs $15.
    - Pros: Foolproof, fun at parties, nostalgic.

    - Cons: Flash so weak it wouldn't illuminate a birthday cake, often results in photos that look like they were taken in a cave.

5. **The Large Format Camera**
    - The final boss of film photography. Requires a tripod, an absurd amount of patience, and a master's degree in physics to operate.
    - Pros: Produces negatives the size of a small tablet, capable of mind-blowing detail.
    - Cons: Requires loading individual sheets of film, making each photo feel like it costs the same as a car payment.

ANALOG REBELLION

Unexpected Quirks of Film Cameras

Every film camera has its own personality. Some cameras work like a dream; others behave like they were designed by a prankster engineer who wanted to see how many people would hurl them into a river. Here are some charming and not-so-charming quirks of film cameras:

- **Light Leaks** – Some cameras refuse to keep light out, which means your perfectly composed shot now has an artsy, accidental sunbeam effect. The more expensive the camera, the less acceptable this is.
- **Sticky Shutters** – Ever take a photo, only to realize the shutter decided to take a lunch break and didn't actually fire? Enjoy your ghostly, double-exposed image of regret.
- **Film Rewind Funhouse** – If you don't rewind your film correctly before opening the back, you will experience the soul-crushing sight of your hard work instantly destroyed by daylight.

- **Mysterious Unexposed Frames** – You shot 36 exposures, but only 30 came out? Congratulations, your camera has secret hiding places you'll never understand.

Ⓐ Ⓝ Ⓐ Ⓛ Ⓞ Ⓖ Ⓡ Ⓔ Ⓑ Ⓔ Ⓛ Ⓛ Ⓘ Ⓞ Ⓝ

Darkroom Development Disasters: When Good Negatives Go Bad

Shooting film isn't just about capturing images—it's about surviving the developing process. Unlike digital photography, where an image is instantly viewable and adjustable, film requires faith, patience, and a willingness to endure catastrophic failures. Every film photographer will eventually find themselves in the darkroom, armed with chemicals, a roll of film, and an overinflated sense of confidence. And that's where things go wrong.

Developing your own film sounds romantic—like you're some artistic genius working in the shadows, pulling masterpieces from chemical baths. In reality, it's mostly standing in the dark, hoping you don't ruin everything in a single misstep. And oh, the ways you can ruin everything:

- **The Coffee Catastrophe** – Some genius on the internet swore you could develop film in coffee. Spoiler: You cannot. Unless you want negatives that smell like burnt regret and look like abstract cave paintings.
- **Fixer First, Regret Later** – Accidentally pouring fixer before the developer is the equivalent of sending your photos into the witness protection program. Nothing survives.
- **The Accidental Thumbprint Collection** – Ever wanted to see what your fingerprints look like embedded into an image of your best friend's wedding? No? Too bad. Welcome to amateur film handling.
- **Light Leak of Doom** – Thought you were being slick when you cracked the developing tank open just a little to check if it was working? Congratulations,

you now have ghostly apparitions floating across your entire roll.

- **The Case of the Missing Images** – You swear you shot 36 frames, but somehow the negatives are blank. Maybe you loaded the film wrong. Maybe you didn't shoot anything at all. Maybe the universe just hates you.

No matter how much you practice, darkroom disasters are a rite of passage. Just embrace the chaos, accept that half your rolls will be tragic, and pretend it was all intentional "experimental art."

ANALOG REBELLION

Why Film Photography Makes You a Better Photographer (And Possibly a Mild Financial Masochist)

Film forces you to slow down. You have to think about lighting, composition, and framing in ways digital photography doesn't demand. You become a more intentional shooter—more patient, more aware of the world around you. Suddenly, you're no longer just snapping pictures—you're making them. Or, at the very least, you're attempting to make them while quietly hemorrhaging money.

Mastering the Exposure Triangle

Let's talk about the exposure triangle—shutter speed, aperture, and ISO—the holy trinity of photography that film shooters learn to worship or perish. With digital, you can brute-force your way through settings like a toddler smashing buttons on a game controller until something works. Film, however, is the strict teacher who slaps your hand with a ruler and says, "No. Figure it out, or your next 36 frames will look like a sad, underexposed crime scene." Master these settings, and you'll unlock the power to properly expose a photo—and sound insufferable at parties ("Well, actually, I shot this at f/8...").

The Beauty of Imperfection

But film's real magic is in its flaws. Light leaks, grain, and accidental double exposures aren't just quirks—they're life lessons in imperfection. Digital lets you delete your mistakes; film forces you to own them. That time you accidentally shot an entire roll at ISO 1600 in broad daylight? Congratulations, you've just created a haunting, high-contrast series that looks like it was filmed on the surface of the sun. Post it on Instagram with the caption "Exploring the fragility of human perception," and watch the likes roll in.

The Financial Toll (and Ultimate Payoff)

Yes, film is expensive. A single roll can cost more than a latte, and developing fees hit like a surprise tax audit. But here's the secret: Film photographers are the ultimate scavengers. We bulk-buy expired film from eBay that smells like a basement, develop it in bathtub chemicals, and call the resulting mold-spore aesthetic "vintage." You'll learn to stretch every dollar like it's the last piece of duct tape holding your camera together.

The Emotional Rollercoaster

And let's not forget the emotional growth. Nothing builds character like finally nailing the perfect shot... only to realize you forgot to take the lens cap off. Or waiting three weeks for your scans, only to discover your masterpiece of a sunset is just 24 photos of your camera bag. Film teaches resilience. Humility. The ability to laugh maniacally while sobbing, "It's fine, I'm fine, this is fine."

The Zen of Film: Why We Keep Doing This to Ourselves

Despite all the quirks, frustrations, and financial ruin, film photography is still a beautiful experience. There's something deeply satisfying about the weight of a real camera, the mechanical click of the shutter, and the anticipation of seeing your photos weeks later. It's an act of patience in an age of instant gratification. It's the photographic equivalent of slow cooking—annoying, but ultimately worth it.

Remember Jerome, the friend who spent an entire afternoon "taking" photos without loading film? When he finally got it right, he told me that the moment he advanced the film and heard the mechanism click into place, it felt like unlocking a secret level in a video game. A small victory, but a victory nonetheless. And when his first actual roll came back with just *one* decent shot—a poorly framed, slightly blurry picture of a streetlamp—he looked at it like it was the Mona Lisa.

That's the magic of film. The struggle makes the reward sweeter. The mistakes make the successes more meaningful. And even when you fail spectacularly, at least you have a great story to tell.

Film isn't dead, and neither is the joy of taking a photo that means something. If you ever want to truly see the world around you, pick up a film camera. Just be prepared to explain to at least one confused bystander why your camera doesn't have a screen.

So go forth. Shoot on film. Make mistakes. And for the love of everything holy, don't open the back of the camera mid-roll.

You'll thank me later.

ANALOG REBELLION

Things I Said While Learning Film Photography

1. **"Wait... I have to MANUALLY focus? What am I, a medieval peasant?"**
2. **"Why does this camera weigh more than my entire sense of responsibility?"**
3. **"I definitely loaded the film... probably... maybe... oh no."**
4. **"How do I check if I have any shots left? *Shakes camera aggressively*"**
5. **"I just opened the back of the camera. I am now officially my own worst enemy."**
6. **"Why does my light meter keep judging me like a disappointed parent?"**

7. **"So 1/250 is fast enough to freeze motion but not my caffeine-induced hand tremors?"**
8. **"I just spent $20 on a roll of film and all I have are blurry regrets."**
9. **"I shot an entire roll thinking it was ISO 800... turns out it was ISO 100, so now everything looks like it was shot through a Syrian dungeon peephole."**
10. **"Why does everyone else's film look like an art gallery, and mine looks like evidence?"**
11. **"That was the best shot of my life—oh wait, I forgot to advance the film."**
12. **"Why does my photo look like it was taken inside the sun?"**
13. **"I spent an hour composing this shot, and it's as out of focus as my life choices."**
14. **"Oh, so *that's* why people use a light meter and not just blind faith."**
15. **"I'll just develop my own film... how hard can it be? (*narrator voice*: It was very hard.)"**
16. **"I now own six cameras, a light meter, three rolls of expired film, and absolutely zero clue what I'm doing."**
17. **"I just paid $40 to develop what looks like an accidental selfie of my thumb."**
18. **"I tried to take a double exposure on purpose, and somehow I made it *worse*."**
19. **"Why did my entire roll come out blank? Oh. I never loaded the film. Fantastic."**
20. **"Despite everything, I still love this stupid, expensive hobby. Someone please stop me."**

Part 2
ANALOG REBELLION AS CULTURE

5

When Fax Machines and Flip Phones Become Counterculture

How Old-School Tech Is Winning Over a Digital Generation

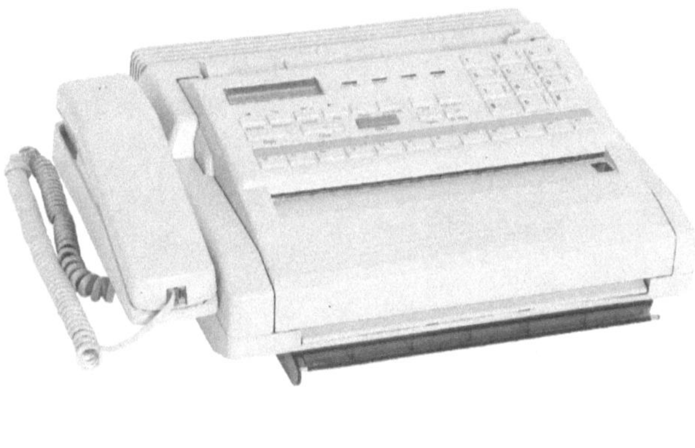

Why the Fax Machine is Back (and Not Just for Boomers)

Gen Z's Love for "Vintage Officecore"

THE OTHER DAY, I walked into a trendy little co-working space that looked like someone had raided a mash-up of a thrift store and a 1980s law office, then threw it all into an open floor plan with overpriced cold brew on tap. There, nestled between a rotary phone and a stack of old IBM punch cards, sat a fax machine. A fax machine. A young woman in high-waisted jeans and wire-rimmed glasses was feeding a sheet of paper into it like she was hand-delivering a message to history itself.

Curious, I asked her why she was using it. She paused for a moment, as if surprised that the answer wasn't already obvious, then adjusted her glasses and said, "Because it works. No failed attachments, no spam filters, no 'sorry, your email bounced' nonsense." She gave the machine a gentle pat, like one might do to a trusty old dog. "Besides, it just feels more... real."

At that moment, I felt like I had stepped into an alternate timeline where fax machines were cool, phone cords were making a comeback, and somewhere, a tech bro was trying to sell people on the blockchain version of a beeper. It was weird, it was wonderful, and it made me realize that we're in the midst of a full-blown office technology renaissance.

A Brief History of the Fax Machine: From Sci-Fi to Office Staple

Believe it or not, the fax machine dates back further than most people think—way further. In fact, it was first conceptualized in the 1840s by Scottish inventor Alexander Bain, who designed a device that could transmit images via electrical signals. Yes, the fax machine is technically older than the telephone.

Let that sink in.

Fast forward to the 20th century, and fax machines began evolving from complex, expensive contraptions into standard office equipment. By the 1960s and '70s, businesses started relying on faxes for fast, reliable document transmission. And then came the 1980s and '90s—the golden age of the fax. If you worked in an office back then, odds are you spent a good chunk of your life standing over a humming, overheating machine, praying that your ten-page report didn't jam halfway through transmission.

For decades, the fax machine was the undisputed king of office communication. Then, the internet arrived, and email gradually replaced faxing as the go-to method for sending documents. By the early 2000s, it seemed like faxes were headed for the technological graveyard alongside floppy disks and dial-up modems.

But here we are, in 2024, watching fax machines stage a comeback like an aging rock band on a nostalgia tour. And this time, it's not just lawyers and doctors keeping them alive—Gen Z is giving these old-school machines a whole new lease on life.

ANALOG REBELLION

Why Gen Z is Bringing Faxes Back

The Unexpected Resurgence of an Outdated Technology

In an era where emails disappear into spam folders, digital signatures get lost in cyberspace, and privacy concerns hover like an uninvited Zoom call, one relic of the past is making a surprising comeback: the fax machine. No, this isn't just a last-ditch effort by technophobic boomers clinging to their landlines. The hum, whir, and screech of a fax connecting over a phone line is being rediscovered by a generation that never had to suffer through dial-up internet—Gen Z.

The Rise of Vintage Officecore

Yes, the same digital natives who grew up with smartphones in their pockets are now embracing analog office equipment in a movement dubbed “Vintage Officecore.” Alongside typewriters, rolodexes, and chunky desktop calculators, fax machines are becoming statement pieces for those who want to blend nostalgia with function. Part rebellion against the sleek, touch-screened efficiency of modern tech, part aesthetic-driven appreciation for the clunky charm of yesteryear, this resurgence isn’t just about looking cool—it’s about taking back control.

For many Gen Z professionals, freelancers, and creatives, the fax machine offers something digital communication often lacks: **tactile, tangible reliability**. In an era where “sent” doesn’t always mean “received,” a fax provides undeniable proof that a document has arrived, inked onto a real sheet of paper. No mysterious “server errors,” no cryptic “undeliverable” messages. Just a comforting confirmation beep and a physical record in hand. Plus, let’s be real—if you send a contract via email, it might disappear into the digital void forever. If you send it via fax, at the very least, someone will have to physically pick it up and wonder, "Who in 2024 is sending a fax?"

Security in a Digital Age

Ironically, Gen Z’s embrace of fax machines isn’t just about nostalgia—it’s also about **privacy**. With cybercrime and data breaches on the rise, many are waking up to the unsettling reality that emails, cloud storage, and digital messaging services aren’t as secure as they seem. A fax, on the other hand, sends information directly from one physical location to another, bypassing the vulnerabilities of the internet. Legal professionals, small business owners, and even some artists are opting for faxes when sending contracts, invoices, and sensitive documents. Plus, let's be honest, hackers aren't exactly sitting around trying to intercept a fax in 2024.

ⒶⓃⒶⓁⓄⒼ ⓇⒺⒷⒺⓁⓁⒾⓄⓃ

Industries That Never Stopped Using Fax Machines

While most of the world moved on to email and cloud-based document sharing, certain industries have remained steadfast in their reliance on fax machines. Hospitals and medical facilities, for instance, still depend heavily on faxing due to strict patient privacy laws like HIPAA, which make email a riskier option. The legal sector is another stronghold of the fax machine—law firms, courts, and government offices still require physical or faxed documents for official filings. Then there's the finance industry, where banks and mortgage companies continue to use faxes to transmit sensitive paperwork securely.

And let's not forget Japan, where the fax machine is practically a national treasure. A 2021 survey found that nearly 34% of Japanese businesses still used fax machines regularly, particularly in industries like government, banking, and healthcare[1]. Despite ongoing efforts to phase them out, many offices remain loyal to faxing, citing its reliability and security. Even the Japanese government, in 2022, struggled to implement a plan to eliminate fax usage, facing strong resistance from bureaucrats who argued that faxing was an essential part of their workflow. Apparently, the quiet dignity of a fax transmission is preferable to the chaos of an overflowing email inbox.

Even niche professions have kept the fax alive. Tattoo parlors use them to send sketches and consent forms. Small-town pharmacies rely on faxes for prescription transfers. Surprisingly, even some artists and designers prefer faxing to scan-and-email setups, citing the aesthetic distortion of a fax printout as an oddly appealing creative touch.

The Fax Machine Resurgence: By the Numbers

Recent research indicates that the fax machine is far from obsolete. In fact, the global fax services market was valued at USD 3.46 billion in 2023 and is projected to reach USD 6.5

---

1. Allen, Jay. "Why Japan's Love for Fax Machines Just Won't Die - Unseen Japan." *Unseen Japan*, 22 Dec. 2021, unseen-japan.com/japan-fax-still-popular. Accessed 7 Feb. 2025.

billion by 2029, growing at a compound annual growth rate (CAGR) of 11.08%[2]. This resurgence is driven by factors such as automation and the integration of fax services into enterprise systems. Additionally, another report found that the global fax services market was valued at USD 3.18 billion in 2022 and is expected to reach USD 5.96 billion by 2028, with an 11.05% CAGR[3].

This unexpected growth is fueled by industries such as healthcare, finance, and law, which continue to rely on faxing for secure document transmission. While many thought the fax machine was destined for the technology graveyard, the numbers suggest otherwise—it's not just nostalgia; it's a functional necessity making a comeback.

Tech Comparison: Fax vs. Email vs. Other Alternatives

In an age of instant communication, where emails travel in seconds and messages disappear into the digital abyss, why would anyone still choose faxing? Let's break it down.

Email

---

2. Research and Markets. "Fax Services Market Focused Insights 2024-2029: Automation and Integration - the next Frontier in Fax Services for Enterprises." *GlobeNewswire News Room*, Research and Markets, 11 Sept. 2024, www.globenewswire.com/news-release/2024/09/11/2944301/28124/en/Fax-Services-Market-Focused-Insights-2024-2029-Automation-and-Integration-The-Next-Frontier-in-Fax-Services-for-Enterprises.html. Accessed 7 Feb. 2025.

3. ---. "Global Fax Services Market Report 2023: A $5.96 Billion Industry by 2028 from $3.18 Billion in 2022, Driven by Replacement of On-Premises with Cloud Fax and Demand from the Healthcare Sector." *GlobeNewswire News Room*, Research and Markets, 20 Feb. 2024, www.globenewswire.com/news-release/2024/02/20/2832032/28124/en/Global-Fax-Services-Market-Report-2023-A-5-96-Billion-Industry-by-2028-from-3-18-Billion-in-2022-Driven-by-Replacement-of-On-Premises-with-Cloud-Fax-and-Demand-from-the-Healthcare-.html. Accessed 7 Feb. 2025.

It's fast, convenient, and often chaotic. Messages get lost in spam folders, buried under a mountain of unread promotions, or flagged for containing a suspiciously large attachment. Ever sent an important document only to have the recipient claim they never received it? Yeah, that happens more often than you'd think.

Cloud-Based Document Sharing

Services like Dropbox, Google Drive, and other cloud storage platforms are fantastic—until they aren't. Files expire, permission settings create an intricate maze of frustration, and convincing a non-tech-savvy client to navigate file-sharing links can feel like explaining quantum physics to a toddler. One wrong click, and poof—your document disappears into the void.

Faxing

You put the paper in, dial the number, and it prints out on the other end. No spam filters, no server failures, no forgotten passwords. It just... works. Plus, that satisfying beep confirming transmission? It's like a little victory soundbite that beats an unreliable email read receipt any day.

Surprisingly, some professionals still argue that faxing is the most secure way to send sensitive documents. Unlike emails, which can be hacked, forwarded, or intercepted, a fax is a direct, point-to-point transmission. This is why banks, law offices, and healthcare facilities continue to swear by them.

And let's not forget the sheer comedic value of telling someone, "I'll fax it to you." Watching their brain short-circuit as they try to remember the last time they saw a fax machine? Priceless.

For some, faxing isn't just a quirky retro habit—it's a statement. Take Oliver, a 26-year-old freelance illustrator who discovered his love for fax machines after a digital nightmare. "I spent hours perfecting a client's design, emailed it, and they claimed they never got it. Twice. Then, I sent it via fax, and boom—instant response. It felt like I had just cracked some ancient code for effective communication."

Then there's Sofia, a small business owner who relies on faxes for contracts and orders. "I used to send everything by email, but things kept getting lost or ending up in junk folders. When I switched to fax, my customers actually started responding faster. Plus, there's something so satisfying about hearing that little confirmation beep—it's like a tiny victory every time."

For Liam, a 23-year-old tech worker with a nostalgic streak, it's all about the aesthetic. "My fax machine is my statement piece. It's absurdly beige, makes ridiculous noises, and looks like it time-traveled from a '90s legal drama. Every time I send a fax, I imagine I'm in an old-school detective movie. I even bought a trench coat to complete the experience."

Or Mia, an artist who stumbled upon a creative use for her fax machine. "I found this ancient-looking fax machine at a flea market and started experimenting with it. The way it distorts images slightly, adds noise to text—it's become part of my art style. Who knew office tech could be a medium?"

And then, of course, there's the aesthetic appeal. "I mean, just look at this thing!" laughs Emily, holding up a vintage beige fax machine she found at a thrift shop. "It's got buttons, rollers, and it makes those ridiculous noises. It's like a pet, but for paperwork."

Despite its reputation as outdated technology, faxing continues to capture the imagination of a new generation. Whether for its reliability, nostalgia, or pure amusement, the fax machine has proven that it's not quite ready to fade into obscurity just yet.

The Ritual and Romance of Faxing

But it's not just about function—there's a romance to the ritual. Feeding a sheet of paper into a machine, hearing the satisfying churn of its rollers, and watching a message physically transmit across the country (or globe) carries a sense of presence and intention missing from a quick email. It's like writing a love letter instead of sending a text that just says "sup?"

And, of course, let's not forget the **aesthetic appeal**. The "Vintage Officecore" trend isn't just about practicality; it's a visual and cultural statement. Picture a desk adorned with a retro beige fax machine, a typewriter on one side, a cup of perfectly sharpened pencils on the other—an antidote to the cold minimalism of the digital age. In the same way that vinyl records and film cameras provide a richer sensory experience than their digital counterparts, old-school office tech offers a way to slow down and engage with the process.

A Countercultural Movement with Purpose

So, what started as a quirky aesthetic choice has turned into a legitimate countercultural movement. Beyond nostalgia and privacy concerns, faxing provides a unique advantage: it ensures secure, verifiable document transmission without the risks of email hacking or cloud service failures. In a world where convenience often comes at the cost of control, fax machines offer a rare blend of simplicity and certainty—qualities that, in an increasingly uncertain digital landscape, are becoming more valuable than ever.

Gen Z isn't just reviving fax machines for the novelty; they're reclaiming them as tools of authenticity, privacy, and self-expression in a world that often feels too fast, too ephemeral, and too exposed. And who knows? Maybe the next time you need to send an important document, you'll find yourself standing in front of one, feeding in a sheet of paper, and thinking, "You know what? This actually makes sense." And if not, at least you'll get to hear that delightful screeching dial tone one more time before it fades into history.

Funny Palate Cleanser: Fax Machine Pickup Lines (Don't Actually Use These)

1. "Are you a fax? Because you're sending all the right signals."
2. "I'd fax you my heart, but I don't want it to jam halfway through."
3. "Are you an office memo? Because I can't stop thinking about you."
4. "You must be a fax machine, because I feel a strong connection."

5. "You're like a fax—old-school but totally reliable."
6. "Can I get your number? I promise to fax you back."
7. "Are you paper jammed? Because you've got me stuck on you."
8. "I'm like a fax machine—I might make weird noises, but I'll always deliver."
9. "Are we on a dedicated line? Because I think we have a secure connection."
10. "I'd fax you a love letter, but I'd rather send it in person."
11. "Do you believe in love at first fax?"
12. "You must be a high-quality printout because you look picture perfect."
13. "Are you a toner cartridge? Because you just filled my heart."
14. "I'd fax you a rose, but technology hasn't caught up with my feelings yet."
15. "Are you a confidential document? Because I want to handle you with care."
16. "You've got me feeling like a long-distance fax—hopelessly in love but slightly delayed."
17. "I'd wait for you like a fax transmission on a slow connection."
18. "Are you a busy signal? Because I can't get through to you."
19. "You must be a thermal fax, because you're heating up my heart."
20. "They say faxing is outdated, but my feelings for you will never be."

6

The Flip Phone Resurgence: Mindfulness or Just Hipster Flex?

Why People Are Ditching Smartphones on Purpose

Once considered relics of a bygone era, flip phones are making an unexpected comeback. Is this a deliberate move toward mindfulness and a rejection of digital overload? Or is it just another hipster trend designed to make answering calls look ironically cool? In an era where screen time is skyrocketing and digital detoxes are becoming mainstream, some people are choosing simplicity over hyper-connectivity—because let's be honest, there's only so much doomscrolling one can do before realizing you've spent two hours reading about the mating habits of deep-sea squids.

There's also the reality that modern smartphones have become digital black holes. One moment, you're checking the weather; the next, you're knee-deep in conspiracy theories about how pigeons are actually government drones. Flip phone users proudly declare themselves immune to the algorithm's endless scroll, free to experience life without the haunting glow of blue light at 2 a.m.

A Brief History of the Flip Phone: From Status Symbol to Obsolete to Reborn

Once upon a time, before touchscreens ruled the world, the flip phone was the pinnacle of mobile cool. The mid-2000s saw a golden era where owning a Motorola Razr meant you were at the height of sophistication, and aggressively snapping it shut after a call was the ultimate power move. Flip phones weren't just communication devices—they were status symbols. Whether you were a business executive, a teenager texting on T9, or Paris Hilton taking blurry selfies, having a flip phone meant you had arrived.

Then came the iPhone, and suddenly, flip phones went from sleek to outdated in record time. As touchscreens took over and mobile internet became a necessity, flip phones were demoted to "burner" status—used primarily by tech-averse grandparents and people who really didn't want to be traced. For over a decade, they lurked in the shadows of smartphone dominance, gathering dust at the bottom of old junk drawers.

Now, against all odds, they're back. But this time, it's not because of cutting-edge technology—it's because people are craving the very thing flip phones once lacked: simplicity.

The Social Consequences of Using a Flip Phone

Using a flip phone in 2025 is an instant conversation starter—or conversation ender, depending on the crowd. Picture this: you're out with friends, and someone whips out the latest smartphone to check directions. You, on the other hand, confidently flip open your relic from 2004. The table goes silent. Someone asks, "Is that...a flip phone?" You nod. Half the group leans in with admiration, the other half looks at you like you just declared you churn your own butter. Either way, you've got their attention.

Friends might assume you're going through an existential crisis. Strangers may glance at you with a mix of nostalgia and pity, unsure whether you're a cutting-edge digital minimalist or just someone who hasn't upgraded since 1998. Meanwhile, tech enthusiasts will ask you why on earth you'd willingly give up touchscreen convenience—right before complaining about their screen time report.

The practical aspects can also get awkward. Good luck explaining to someone that you can't scan the QR code menu at a restaurant, and no, you don't have Google Maps—so maybe we should just ask for directions like it's 1995. And don't even try taking a group selfie. Your friends will take one look at your 1.3-megapixel camera and suggest, politely, that maybe someone else should handle it.

Yet, despite these minor inconveniences, flip phone users revel in the simplicity. No endless notifications, no unnecessary apps, no mindless scrolling—just a functional device that forces you to be present. And when someone asks for your number, flipping it open with a dramatic snap is infinitely cooler than fumbling with Face ID.

In a world where hyper-connectivity reigns supreme, choosing a flip phone isn't just a practical decision—it's a statement. Whether for mindfulness, rebellion, or just an appreciation for the satisfying *snap* of a well-built hinge, the resurgence of flip phones proves that sometimes, simpler really is better. To truly understand the flip phone resurgence, let's take a moment to appreciate what it's like to live with one. Meet Chandler, a 29-year-old ice-cream store owner who, after one too many doomscrolling sessions and social media-induced existential crises, decided to switch back to a flip phone. "I was spending hours a day staring at my phone, achieving nothing but a deep knowledge of which celebrities liked each other's Instagram posts. It was like a black hole for productivity."

At first, the transition was rough. "I kept instinctively swiping on the screen, but of course, nothing happened. I also forgot how texting worked—it took me 45 seconds to type 'okay.' But after a while, I started noticing things. I actually paid attention to my surroundings. I didn't default to Googling every random thought that popped into my head. And for the first time in years, I actually finished reading a book."

Chandler's experience mirrors that of many who make the switch. The absence of endless scrolling leads to unexpected benefits: better focus, improved sleep, and a sense of relief from the constant digital noise. And yes, the novelty of snapping the phone shut after a call remains immensely satisfying.

ANALOG REBELLION

Flip Phones in Pop Culture: From Action Heroes to Y2K Icons

Before they became a symbol of digital minimalism, flip phones were a staple of pop culture. In action movies, spies and criminals couldn't get through a single scene without flipping open a burner phone, making a cryptic call, then dramatically snapping it shut and tossing it into the nearest trash can.

Think *The Matrix*—Neo receives a sleek black flip phone in a mysterious package, signaling the start of his journey into a hidden reality. Or *24*, where Jack Bauer seemingly carried an endless supply of disposable flip phones, each one lasting precisely until it was no longer useful.

But it wasn't just action heroes. In the early 2000s, flip phones were the ultimate status symbol. Every teen drama featured a character aggressively texting on a Razr while ignoring their parents. Paris Hilton practically had one surgically attached to her hand. Even reality TV icons treated their phones as accessories, custom bedazzling them like some kind of telecommunications tiara.

Fast forward to today, and those same nostalgic associations make flip phones even more appealing. Nostalgia isn't just about reminiscing—it's a powerful marketing tool. Companies have long tapped into the emotional pull of the past to sell everything from vinyl records to vintage sneakers, and flip phones are no exception. Research shows that consumers are drawn to products that remind them of a simpler time, a phenomenon often referred to as "retro branding." In a world of overstimulation, the flip phone offers a comforting throwback, a tangible link to an era before notifications and algorithm-driven distractions dictated daily life.

ANALOG REBELLION

The Appeal of the Flip Phone

For many, the flip phone represents more than just nostalgia—it's a lifestyle shift. Users report feeling more present, less anxious, and more engaged with the world around them. Without the constant barrage of notifications, doomscrolling, and algorithm-driven distractions, flip phone users claim they regain control of their time and attention. Plus, there's something undeniably satisfying about snapping a phone shut with dramatic flair. Try slamming an iPhone shut after a heated conversation—you'll just end up looking like a frustrated toddler tapping on a toy.

Even social interactions become different with a flip phone. Instead of the dead-eyed stare of someone half-listening while scrolling Instagram, flip phone users actually make eye contact. Shocking, I know. Friends and family may initially be baffled by the decision, but there's no denying the impact of being fully present. Also, if someone asks you to "Google something real quick," you have a built-in excuse to just sit there and enjoy the mystery.

As more people seek ways to unplug and reclaim their time, the humble flip phone is proving that sometimes, the simplest technology is the most revolutionary.

Mindfulness Over Mindlessness

Digital well-being advocates argue that returning to a simpler device fosters mindfulness. Instead of instinctively reaching for a phone to check social media, flip phone users engage more with real-world experiences. Conversations become deeper, boredom sparks creativity, and relationships strengthen without the ever-present distraction of endless apps. And best of all, nobody gets stuck in the existential crisis of figuring out what to watch on Netflix for an hour before deciding to rewatch *The Office* for the twelfth time.

There's also a noticeable decline in tech-related anxiety. A 2021 study published in *Computers in Human Behavior*[1] found that reducing smartphone usage led to lower stress levels, improved sleep quality, and higher overall life satisfaction. Many flip phone users report feeling a greater sense of control over their time and attention, free from the constant pings and dopamine-driven distractions of modern devices. No more obsessing over battery percentages, unread notifications, or that one mysterious app update that changes everything for no reason. Flip phone users report lower stress levels, better sleep, and—perhaps most impressively—the ability to sit in silence without twitching.

ⒶⓃⒶⓁⓄⒼ ⓇⒺⒷⒺⓁⓁⒾⓄⓃ

The Hipster Factor: Rebellion in a Tech-Obsessed Age

Flip Phones: Drug Lord vs. Hipster Edition

Flip phones have long been the communication device of choice for two very different groups: international criminals and urban hipsters. Here's how they stack up:

> **Drug Lord:** Uses a flip phone to avoid government surveillance.

1. ramjan, lucie M. "The Negative Impact of Smartphone Usage on Nursing Students: An Integrative Literature Review." *Nurse Education Today*, vol. 102, 1 July 2021, p. 104909, www.sciencedirect.com/science/article/abs/pii/S0260691721001660, https://doi.org/10.1016/j.nedt.2021.104909.

Hipster: Uses a flip phone to avoid Instagram.

Terrorist in an Action Movie: Tosses the phone in a dumpster after making a cryptic call.

Hipster: Tosses the phone in their tote bag after texting "u up?" on T9.

Spy on the Run: Smashes phone in half to cover their tracks.

Hipster: Accidentally sits on phone; it survives without a scratch.

Cartel Boss: Uses multiple flip phones to run an international empire.

Hipster: Uses multiple flip phones as a retro fashion statement.

Mafia Underboss: Answers with a curt "It's done."

Hipster: Answers with "Hey, I found this rad hole-in-the-wall café. Wanna go?"

Villain in a Jason Bourne Movie: Stares menacingly at flip phone before making a call.

Hipster: Stares at flip phone wondering if their date will think it's quirky or just weird.

Of course, it wouldn't be a cultural shift without the hipsters. For some, adopting a flip phone is less about mindfulness and more about standing out. In a world where cutting-edge smartphones dominate, flipping open a clunky old device is a statement—an intentional step away from the mainstream. It signals a rejection of Big Tech, planned obsolescence, and the relentless upgrade cycle. Because nothing screams *I'm different* quite like answering a call on a device last seen in a 2005 episode of *Laguna Beach.*

ⒶⓃⒶⓁⓄⒼ ⓇⒺⒷⒺⓁⓁⒾⓄⓃ

The Role of Tech Companies in the Resurgence

Some tech companies never truly abandoned the flip phone, especially in niche markets. Brands like Samsung and Motorola have been quietly churning out new models, particularly for older users and business professionals who prefer simplicity over smartphone chaos. But recently, major phone manufacturers have leaned into nostalgia, launching modern flip-style smartphones that blend old-school functionality with cutting-edge technology.

Take the Samsung Galaxy Z Flip—an attempt to bridge the gap between flip phone purists and touchscreen addicts. It folds in half like its predecessors, but instead of a simple LED screen with a monochrome display, it boasts a foldable touchscreen, high-end cameras, and app compatibility. Some argue that these modern takes on the flip phone defeat the purpose of a digital detox, but their success proves that the market for compact, foldable devices is alive and well.

Meanwhile, minimalist phone companies such as Light Phone and Punkt have introduced deliberately stripped-down devices that mimic the simplicity of classic flip phones while ditching social media and most internet functions. These cater directly to those seeking a break from digital overload without resorting to an outdated relic from 2005.

So, while old-school flip phones are making a comeback for some, tech companies are ensuring that even in a world of nostalgia, there's always an upgrade available.

ⒶⓃⒶⓁⓄⒼ ⓇⒺⒷⒺⓁⓁⒾⓄⓃ

Practical Benefits: More Than Just Aesthetic

Beyond mindfulness and rebellion, there are practical reasons for ditching smartphones. Flip phones are significantly cheaper than their touchscreen counterparts, offering a budget-friendly alternative without the pressure of yearly upgrades. They also generate less electronic waste, making them a more sustainable choice for those concerned about the environmental impact of rapid tech turnover. Plus, they're nearly indestructible—drop a flip phone, and it laughs in the face of gravity, unlike a smartphone that shatters at the mere thought of a hard surface.

Flip phones also eliminate digital clutter. No app hoarding, no "storage almost full" warnings, and no endless customization options that lead to three-hour rabbit holes of wallpaper selection. They force users to interact with their devices in a straightforward, no-frills manner, which—ironically—feels like a breath of fresh air.

ⒶⓃⒶⓁⓄⒼ ⓇⒺⒷⒺⓁⓁⒾⓄⓃ

Is the Future Flip?

Flip phones may have started as a niche movement, but they are increasingly becoming a legitimate challenger to smartphone dominance. Major tech companies have taken notice, with brands like Samsung and Motorola releasing modern flip-style devices that blend nostalgia with cutting-edge features. Industry analysts predict that as digital detox trends continue, demand for simplified mobile experiences will grow, ensuring that the flip phone resurgence isn't just a passing fad but a lasting shift in consumer preferences.

While tech companies continue to push faster, smarter devices, a counterculture of intentional disconnection is growing. Whether for mindfulness, rebellion, or practicality, the flip phone resurgence proves that sometimes, simpler really is better. After all, it's hard to get lost in the digital abyss when your phone's biggest feature is the ability to make a call and, if you're feeling really fancy, send a text—provided you have the patience to T9 it out.

So, is the future flip? Maybe not entirely. But as long as people keep reaching for simpler alternatives to a world drowning in notifications, the humble flip phone will continue to serve as a pocket-sized act of defiance—one slow, painstaking text at a time.

7

Analog Underground

Secret Societies of Letter Writers and Tape Traders

Between the cracks of our hyper-connected digital world, a thriving underground network of analog devotees is keeping the art of slow communication alive. These individuals swap handwritten letters, carefully curated mixtapes, and lovingly crafted zines as a form of quiet rebellion against instant gratification. I stumbled into this secret society quite by accident when a mysterious, hand-addressed envelope landed in my mailbox. Inside was a beautifully typed invitation to join something called *The Midnight Typists' Collective.*

This invitation marked my initiation into one of the most fascinating subcultures I've encountered—an analog movement where the art of letter writing isn't just surviving; it's thriving in the shadows of our digital age. These aren't just your grandmother's pen pals (though, intriguingly, some grandmothers are key players in these networks, and honestly, I wouldn't mess with them). Instead, they are vibrant communities of artists, musicians, and writers who believe that sometimes, the best way to push back against digital monotony is with a sheet of paper and a stamp—because nothing says "rebellion" like struggling to find a post office open past 5 p.m.

The Social Hubs of the Analog World

In basements, coffee shops, and converted garages across the country, small groups gather to trade handcrafted zines, exchange mixtapes that rival the best DJ sets, and share letters that feel more like works of art than simple correspondence. It's a secret society for those who think digital convenience has drained communication of its soul. It's a speakeasy for people who believe Microsoft Word's handwriting font is a personal affront to humanity.

The level of dedication in these communities is both inspiring and, at times, intimidating. I've met people who can identify different types of paper stock by touch alone, like sommeliers sniffing out notes of oak and blackberry in a fine wine. Some own more typewriters than they do digital devices, and others hoard stamps the way tech enthusiasts hoard gadgets. And if you ever want to see true horror, tell a zine-maker that you accidentally printed their masterpiece on 20-pound office paper—then prepare for the kind of reaction normally reserved for someone who microwaves fish in an office break room.

Some clubs, like the Handwritten Letter Appreciation Society[1] even claim to have 1m803 members in 38 countries!

Operating Under the Digital Radar

What makes these underground analog networks so compelling is how they continue to grow while flying under the radar of mainstream digital culture. They are like vinyl-loving ninjas, slipping between the cracks of a world obsessed with algorithms and engagement metrics. And here's the delicious irony—many of these so-called digital refuseniks first discover their communities through social media hashtags and online forums. Yet, once they find their people, they unplug, opting for tangible connection over digital distraction. It's the modern-day equivalent of finding enlightenment through an Instagram reel about how to delete Instagram.

ⒶⓃⒶⓁⓄⒼ ⓇⒺⒷⒺⓁⓁⒾⓄⓃ

The Future of the Analog Rebellion

As technology continues its relentless march forward, these analog enthusiasts are proving that not everything needs to move at the speed of light. Whether through the scratch of a fountain pen, the satisfying *click-clack* of a typewriter, or the hiss of a cassette tape, they remind us that some experiences are meant to be slow, deliberate, and beautifully imperfect. In a world that constantly demands our attention, perhaps the most radical act of all is to simply put down our phones and pick up a piece of paper. Or better yet, send a letter—because nothing terrifies a digital native quite like an envelope with their name on it that *isn't* a bill or jury duty notice.

ⒶⓃⒶⓁⓄⒼ ⓇⒺⒷⒺⓁⓁⒾⓄⓃ

Modern Pen Pal Networks: The Revival of Letter Writing

The modern pen pal movement isn't just your grandmother's hobby anymore—though if she's part of it, she's probably more hardcore than you.—though I did recently catch my grandmother trading artisanal washi tape with a 19-year-old letter enthusiast from Portland. These days, letter-writing communities are popping up faster than overpriced oat milk lattes in Brooklyn., each with their own peculiar rules and rituals that would make Emily Post spin in her carefully addressed grave.

Part of the resurgence is driven by digital fatigue, as more people crave deeper, more meaningful connections in a world where communication has been reduced to emojis and double-taps. But let's not underestimate the sheer joy of getting actual mail that isn't a bill or an unsolicited coupon for discounted gutter cleaning.

I recently joined one such group called *The Snail Mail Society* (because apparently, we analog folks can't resist a good pun). Their monthly meetings are like Fight Club for stationery addicts—the first rule is you don't talk about your hand cramps. The second rule is you absolutely must have at least three different types of sealing wax. I showed up to my first meeting with a single gold sealing wax stick and received looks that made me feel like I'd brought a crayon to an art gallery.

What fascinates me most about these modern letter-writing networks is how they've managed to create a perfect paradox—using Instagram to find people who want to avoid using Instagram to communicate. It's like using a dating app to find someone who hates technology. These communities have their own hashtags, their own social media groups, and their own websites—all dedicated to helping people disconnect from... well, social media groups and websites.

The dedication to the craft is both impressive and slightly concerning. I've watched members spend hours perfecting their wax seals, only to realize they forgot to put the letter inside the envelope. One particularly memorable evening, a new member accidentally set fire to her vintage lace handkerchief while trying to achieve the perfect wax drip pattern. The room's collective gasp was followed by someone shouting, "Quick! Document this for the analog disaster Instagram!"

ANALOG REBELLION

Why People Are Choosing Slow Communication

Beyond the quirky rituals and occasional pyrotechnics, these networks are creating something remarkable. In an age where most of our communication is reduced to emojis and "likes," these letter writers are crafting deeply personal, tangible connections. Each envelope that arrives in your mailbox is like a tiny time capsule, carrying not just words but physical evidence of someone's thoughts and efforts—complete with coffee stains, smudged ink, and the occasional cat paw print (which, I'm told, adds character).

The most surprising aspect of this revival isn't just who's participating—it's how they're reimagining the whole concept of pen pal networks. Traditional one-to-one correspondence has evolved into elaborate web-like communities, where letters might pass through multiple hands, collecting additions and annotations along the way. Think of it as an analog version of a group chat, but one that takes weeks to unfold and requires significantly more trips to the post office.

I've watched teenagers spend hours crafting elaborate mail art, turning simple envelopes into miniature galleries. There's something wonderfully defiant about watching a generation raised on instant messaging choosing to wait weeks for a response. It's like they've discovered a secret superpower—patience. Though I suspect part of the appeal is that it's much harder for their parents to monitor their mail than their text messages.

The beauty of these modern pen pal networks lies in their contradiction—they're simultaneously a rebellion against and a product of our digital age. They've taken the best parts of social networking—the ability to connect with like-minded people across vast distances—and combined it with the intimacy and intentionality of handwritten correspondence. It's slow social media, where every 'post' is handwritten, every 'comment' requires postage, and every 'like' is the implicit satisfaction of knowing your letter didn't end up in the neighbor's mailbox.

And yes, sometimes letters get lost, ink smudges render words illegible, and international mail takes so long you begin to question if you're actually corresponding with someone or just shouting into a very well-decorated void. But that's part of the charm. In a world where everything is instant, trackable, and editable, there's something refreshingly human about these imperfect, unpredictable connections.

ⒶⓃⒶⓁⓄⒼ ⓇⒺⒷⒺⓁⓁⒾⓄⓃ

Underground Tape Trading: From Mixtapes to Music Discovery

If you think mixtape trading died with the 90s, you clearly haven't stumbled into the fascinating underground world of modern tape enthusiasts. These aren't your average nostalgia seekers—they're sonic archaeologists, musical matchmakers, and analog audio evangelists all rolled into one caffeine-fueled community. And let me tell you, they take their tape trading more seriously than a cat takes its napping schedule—except these folks don't even get distracted by a sunbeam halfway through.

I discovered this subculture last year when I found myself in what can only be described as a 'tape dealer's den'—actually just some guy named Steve's basement, but he had more cassettes than my local library had books. The walls were lined with meticulously labeled shelves, each tape adorned with hand-drawn artwork that would make a graphic designer weep. Steve, sporting a t-shirt that read "Still Rolling Since '85," proceeded to give me a crash course in modern tape trading etiquette. Rule number one: never, ever fast-forward through someone's carefully curated mix. Apparently, that's the tape trading equivalent of skipping to the end of a book.

What makes today's tape trading scene particularly fascinating is how it's evolved from its underground roots. These aren't just people sharing mainstream music—they're creating elaborate audio time capsules. I recently received a tape that included field recordings of New York City subway performers, mixed with obscure Japanese jazz tracks, seamlessly transitioning into a recording of someone's grandmother telling stories about the 60s. Try finding that on Spotify's algorithm—unless they've introduced a 'hipster cryptic chaos' playlist that I'm unaware of.

The most surprising aspect of this revival isn't just who's participating—it's how they're using modern technology to facilitate their analog obsession. There are entire Discord servers dedicated to coordinating tape trades, where members spend hours debating the merits of different tape brands and sharing tips on how to restore vintage Walkmans. It's like a speakeasy for people who think digital audio is too mainstream, except instead of bootleg gin, they're trading bootleg recordings of that one concert where the singer sneezed mid-chorus.

The dedication to the craft is both impressive and slightly concerning. I've watched traders spend entire weekends perfecting the transitions between songs, treating dead air like a personal insult to their ancestors. One particularly memorable evening, I witnessed someone nearly have an existential crisis trying to decide if they should include both versions of a rare B-side on their mix. The tension in the room was thicker than a tangled cassette tape in a car stereo on a 100-degree day—right before it gets eaten alive.

But beyond the obsessive attention to detail and occasional tape deck meltdowns, there's something beautifully human about this whole enterprise. In an age where algorithms curate our musical experiences, these analog audiophiles are creating something deeply personal. Each tape is like a handcrafted letter, except instead of words, it speaks in carefully chosen songs and intentionally placed pauses.

The Rituals and Superstitions of Tape Traders

The community has its own peculiar rituals and superstitions. Some traders insist on recording only during specific phases of the moon (something about magnetic fields and audio quality—I'm not entirely convinced). Others have elaborate ceremonies involving incense and vintage audio equipment that would make a vinyl collector look positively casual in comparison. I once watched someone perform what they called a "tape cleansing ritual" before recording, which mostly involved waving the cassette over a running tape deck while mumbling something about Dolby noise reduction.

What's particularly ironic about this whole scene is how the digital world has actually helped preserve and evolve these analog practices. Social media has become the unlikely matchmaker for tape enthusiasts, connecting people who still believe the hiss and wow of magnetic tape adds character to music. It's like using a dating app to find someone who insists on courting you via medieval scrolls delivered by carrier pigeon.

The Future of Mixtape Culture

The beauty of modern tape trading lies in its beautiful contradiction—it's simultaneously a rebellion against and a celebration of how we consume music. These traders have taken the best parts of music sharing—the discovery, the personal connection, the surprise of finding something new—and combined it with the intentionality and effort that only analog technology can demand. It's slow listening, if you will, where every track is chosen with purpose and every pause is part of the performance.

Whether mixtape culture remains a niche passion or expands into a larger movement, one thing is clear: there will always be people who prefer their music with a little more personality and a lot more hiss. And as long as there are cassettes to be recorded and traded, there will be tape traders willing to carefully craft the perfect mix—one spliced tape loop at a time.

ANALOG REBELLION

Zine Culture: Self-Published Rebellion in the Digital Age

In an era where anyone with a smartphone can blast their thoughts to millions with a single tap, there's something beautifully defiant about people hunching over photocopiers at midnight, manually collating pages of their handcrafted zines. It's like watching medieval monks painstakingly illuminate manuscripts, except these modern scribes are wielding glue sticks, wrestling with staplers, and passionately debating the merits of cardstock thickness like it's a life-or-death decision.

I recently stumbled into the world of modern zine-making when I attended what I thought was a regular art show, only to discover an underground zine fair buzzing with the same DIY spirit that once fueled the punk and riot grrrl scenes. While classic zines were often raw, cut-and-paste manifestos passed between subcultures, today's zine creators blend old-school aesthetics with modern themes, tackling everything from digital burnout to niche internet humor. The format has evolved, but the heart of zine culture—rebellion, self-expression, and a deep disdain for mass-produced content—remains the same.

The first thing that struck me was the sheer variety—meticulously hand-drawn comics detailing someone's existential crisis at the laundromat, political manifestos typed on vintage typewriters with margins so wonky they'd give a graphic designer heart palpitations, and one zine that was literally just a catalog of things the creator had found in their pockets, complete with authentic lint samples taped inside. One particularly memorable zine was simply titled *Things I Found in My Pocket*, which was exactly what it advertised, complete with actual lint carefully taped to each page.

The Beauty of Imperfection

The beauty of zine culture lies in its deliberate rejection of digital perfection. In a world where we can endlessly edit and filter our content, zines embrace the smudged ink, the slightly crooked staples, and the occasional coffee stain that somehow makes it into every copy. It's like watching a rebellion against spell-check and auto-correct, one photocopied page at a time.

What fascinates me most about modern zine culture is how it's managed to evolve without selling out—like punk rock, but with better typography and fewer smashed guitars. These aren't just photocopied newsletters anymore; they're intricate works of art that combine traditional cut-and-paste techniques with elements that deliberately mock digital culture. I recently received a zine that was entirely about the anxiety of choosing the perfect emoji, illustrated entirely in hand-drawn pictographs that looked like they'd been sketched by someone having an existential crisis.

The Community of Zine Makers

The community aspect of zine-making is perhaps its most charming feature. At a recent zine swap meet, I watched as a retired librarian traded her meticulously researched zine about local bird species with a teenager whose publication consisted entirely of photographs of vintage keyboards arranged to look like faces. The beauty of it was that both creators treated each other's work with equal reverence—in the zine world, there's no algorithm to determine what gets seen and what doesn't.

The process of making a zine is like a combination of meditation and mild self-inflicted torture. One moment, you're blissfully collaging, lost in the joy of creation; the next, you've stapled your thumb to page three and are questioning all your life choices. Last month, I decided to make my own zine about my adventures in analog technology. What started as a simple project turned into a three-day odyssey involving two paper cuts, one jammed photocopier, and an unfortunate incident with a stapler that I'm still not ready to talk about. But there was something incredibly satisfying about physically assembling each copy, knowing that every slight imperfection made each one unique.

The Rituals and Traditions of Zine Culture

Modern zine makers have developed their own peculiar rituals and traditions. There's an unspoken rule that every zine fair must have at least one person who brings a typewriter just to dramatically compose spontaneous poetry for visitors. Another tradition involves the *zine trade chain* where each creator must trade with someone they've never met before, leading to some of the most wonderfully random cultural exchanges I've ever witnessed.

What's particularly ironic about contemporary zine culture is how it uses digital platforms to facilitate analog distribution. There are entire Instagram accounts dedicated to showcasing zine covers, though getting your hands on an actual copy still requires good old-fashioned snail mail or meeting someone in person at a zine fair. It's like using a dating app to find someone to exchange handwritten letters with—a perfect example of how analog and digital cultures can coexist rather than compete.

Why Zines Matter in a Digital World

The physical nature of zines creates a unique form of accountability that's missing from digital publishing. When you're making physical copies of something, you tend to think more carefully about what you're putting out into the world. After all, you can't just hit delete if you change your mind—those photocopied manifestos about your ex's terrible taste in music are out there forever, floating around in the analog ether.

Perhaps what makes zine culture so appealing in our digital age is its emphasis on human connection through imperfection. Every smudged page, every slightly off-center image, every hand-drawn correction becomes part of the story. In a world where we're constantly curating perfect digital versions of ourselves, there's something revolutionary about creating something that proudly wears its flaws on its photocopied sleeve.

In a world where our phones buzz with endless notifications and our inboxes overflow with digital noise, there's something wonderfully subversive about these analog underground networks. They're like secret societies for people who think the best form of encryption is terrible handwriting. As I've discovered through my adventures in these communities—from late-night mixtape recording sessions to zine-making marathons that left my fingers permanently stained with ink—there's an undeniable magic in creating something you can actually hold.

ANALOG REBELLION

The Future of Analog Creation

What makes these analog communities truly special isn't just their dedication to physical media; it's their ability to create genuine human connections in an increasingly virtual world. These aren't just hobbyists clutching onto the past; they're revolutionaries who've figured out that sometimes the best way to move forward is to take a step back. They're proving that in an age where everything is instant and disposable, there's profound value in taking your time and making something that might outlast your latest phone upgrade.

The beautiful irony of this analog underground isn't lost on me—using Instagram to find people who reject Instagram, joining Discord servers to discuss the joys of disconnecting, and hunting for zine fairs via Google Maps instead of a cryptic hand-drawn flyer taped to a lamppost like the good old days. Here we are, using Instagram to find people who want to write letters or Discord to organize tape trading meetups. It's like using a dating app to find someone who believes in arranged marriages. But maybe that's the point—these communities aren't about rejecting the digital world entirely; they're about finding a better balance, about creating spaces where we can slow down and remember what it feels like to make something with our hands.

So maybe the next time you're feeling overwhelmed by the digital deluge, remember there's probably a secret society of analog enthusiasts meeting in a coffee shop basement near you. Just look for the people with ink-stained fingers and suspicious bulges in their bags that are probably full of mixtapes. They might just show you that the best kind of social network doesn't require a password—just a stamp and a little patience.

8

The Social Media Paradox

Why We Post About Our Offline Lives

A Brief Historical Perspective

THERE WAS A TIME when sharing memories meant pulling out a dusty photo album and forcing your relatives to feign interest. They'd sit politely, nodding at slightly out-of-focus pictures of family vacations and birthday parties, occasionally offering a "Wow, you were so young there!" just to prove they were paying attention. Now, we meticulously document our lattes for strangers on the internet, ensuring the perfect angle, lighting, and caption before taking a single sip.

"Golden hour light, ethically sourced oat milk, a sprinkle of cinnamon, and a reminder to romanticize your life #LatteArt #SlowLiving"—because nothing says 'living in the moment' like spending ten minutes crafting a caption about it.

In the pre-social media days, people kept handwritten journals, filled with deeply personal thoughts and reflections that no one else was meant to see. Today, we call those "journals" Instagram captions, and we craft them with the same care once reserved for love letters. The irony? The love letters were private, but our online diary entries are for an audience, complete with hashtags for maximum engagement.

The Contradiction of Analog on Instagram

A few months ago, I decided to embrace the analog life. No more screens, no more mindless scrolling, just me, my thoughts, and the simple pleasures of the real world. Naturally, my first instinct was to announce this revelation... on social media. "Going off the grid for a while! " I typed, making sure to attach an aesthetically pleasing photo of a typewriter next to a steaming cup of herbal tea. The irony was lost on absolutely no one—except me, for about five minutes.

Then, as if on cue, the panic set in. What if someone tried to reach me? What if I missed an important update? What if something major happened and I wasn't there to witness it in real-time? My commitment to digital minimalism lasted all of twelve hours before I found myself "just checking" Instagram. For the record, nothing had happened. The world was fine. But someone had posted a picture of their sourdough starter, and suddenly I was all in.

Nothing quite captures the irony of modern life like seeing someone post an artfully staged Polaroid with the caption: "Unplugging for the weekend!" Right. Because nothing says "living in the moment" like spending twenty minutes adjusting the lighting, slapping on the perfect vintage filter, and coming up with a hashtag that screams *effortless authenticity.*

We live in an age where people spend more time documenting an experience than actually experiencing it. Take a scenic hike, for example. Gone are the days when one simply enjoyed the crisp mountain air. Now, it's a multi-step production: find the best viewpoint, strike a pose that looks natural but is actually staged, take 47 versions of the same shot, and then spend the next hour agonizing over whether #naturevibes or #wildandfree will get more engagement. Heaven forbid you just take it all in and let the memory live in your brain instead of the cloud.

Let's be honest—if a tree falls in the forest and no one Instagrams it, did it even fall? And if it did, wouldn't it be better with a #nofilter caption and a perfectly timed golden hour glow? After all, authenticity is only authentic if it gets engagement. The rise of analog nostalgia—film cameras, vinyl records, handwritten letters—has been accompanied by an almost obsessive need to document it online. We don't just experience moments anymore; we curate them. And then we painstakingly craft a caption about how little we care about curating them. The cycle is as exhausting as it is hilarious.

It's the ultimate contradiction: the people most passionate about analog living are the same ones who ensure that everyone knows about it—digitally. A handwritten letter has more impact when photographed artfully next to a cappuccino. A film camera's grainy beauty is incomplete unless posted in a carousel of images captioned, "Some moments don't need a screen... except this one."

A Guide to the Perfect Digital Detox Announcement

So, you've decided to take a break from social media. Congratulations! But before you unplug, there's one crucial step you can't ignore—announcing your departure. After all, how will people know you're offline if you don't post about it first?

Step 1: The Grand Announcement

Make sure your final post before detoxing is a deep, introspective quote from Rumi or Thoreau. Bonus points if it's paired with an aesthetic shot of a foggy mountain or a steaming cup of tea next to an open book. Caption options include:

- "Sometimes, the best connection is no connection at all. "
- "Logging off to log in to real life. See you when I see you. "
- "Going where the WiFi is weak and the soul is strong."

Step 2: The Strategic Goodbye

Disable notifications... but keep checking them, just in case. What if someone comments something insightful? What if people DM you in panic, wondering how they'll survive without your daily posts? You wouldn't want to leave them hanging.

Step 3: The Temptation Phase

You're officially detoxing. Time to delete the apps from your phone... but keep them easily re-downloadable, you know, just in case you need to "check one quick thing." Maybe you'll casually borrow a friend's phone to take a peek at what's trending. Not posting isn't the same as not looking, right?

Step 4: The Triumphant Return

You're back, baby! And you've *grown*. Of course, this requires a reflective comeback post—something vague yet profound. Include a nature shot and wax poetic about how "eye-opening" your time away was. Classic caption ideas:

- "Time away reminded me how beautiful life is beyond the screen. But also, hi again."
- "Feeling refreshed, recharged, and ready to overshare once more."
- "Absence makes the likes grow fonder. What did I miss?"

With this foolproof guide, you'll be ready to execute the perfect detox announcement while still basking in the digital validation you were trying to escape. And if you ever need to detox again? Don't worry—you can always repost your farewell message.

There's a special kind of cognitive dissonance in announcing to your followers that you're taking a break from social media—on social media. "Logging off for a while, need to clear my mind ," they type, before aggressively refreshing their notifications for the next two hours to see if anyone noticed their absence.

The modern digital detox has become a performance piece. It's no longer enough to simply take a break from social media—you have to tell the world you're doing it, otherwise how will they know? It's the equivalent of holding a press conference to announce that you're going into hiding.

The truth is, most digital detoxes last about as long as a New Year's resolution at an all-you-can-eat buffet. One minute, we're waxing poetic about how much better life is without screens; the next, we're sneaking onto Twitter to see if the world collapsed in our absence (spoiler: it didn't, but someone did make a meme about it).

And then there's the post-detox return, which comes with its own unwritten rules. One does not simply reappear on social media like nothing happened. No, there must be a grand re-entrance. A well-crafted post that signals wisdom gained from time away, ideally accompanied by a black-and-white photo of a tree or a moody sunset. Bonus points if you include an abstract quote about 'being present'—ironically typed on your phone while double-checking the engagement stats every five minutes. No, there must be a reflective post—something vague yet profound, usually accompanied by a nature photo, detailing the "perspective" gained from time away. Bonus points if the caption includes a quote about being present, despite the fact that the person posting it has already checked the engagement stats five times.

ANALOG REBELLION

The Role of Influencers & Brands

Of course, where there's a trend, there's a marketing opportunity. Even major brands have gotten in on the "offline" aesthetic, selling 'minimalist' planners, detox retreats, and handcrafted journals—because nothing says escaping capitalism like spending $45 on a notebook to write about escaping capitalism.

Influencers, naturally, have also found a way to monetize digital detoxing. They film "unplugging" vlogs with carefully edited shots of themselves meditating in scenic locations, only to upload them from a five-star eco-resort with premium WiFi. Their message? Step away from screens... after you use a screen to like and subscribe.

Companies push products designed to help you "disconnect," which, ironically, you learn about through targeted ads while scrolling your phone. There's even a $150 "digital detox candle" that promises to cleanse you of screen addiction—presumably by setting your WiFi router on fire. And let's not forget the "anti-tech" retreats that charge you the price of a new laptop just to sit in a yurt and stare at a rock for a weekend. Whether it's a $300 meditation retreat that promises to free you from your notifications or an app that reminds you to stop using apps, the paradox remains: the industry of unplugging is, in itself, plugged in.

Why We Can't Help Ourselves

It's not that we don't want to disconnect—it's just that we want everyone to know we're disconnecting. Because what's the point of self-improvement if there isn't an audience to validate it? And therein lies the paradox: We crave real-life experiences, but we also crave the applause for having them. We want to escape the algorithm, but we also kind of want the algorithm to notice we escaped.

And let's not forget the fear of missing out—not on world events, but on our *own* lives. If we don't document it, did it even happen? How will we remember the blissful simplicity of a screen-free Sunday if there isn't a post with soft lighting and a wistful caption saying, "Sundays are for slowing down"? The irony is lost on exactly no one—except, apparently, the person posting it.

So, next time you see a post about someone "embracing the analog life," just remember—somewhere, right before they hit upload, there was a frantic search for WiFi, a deep existential debate about which sepia filter felt the most *authentic*, and at least five minutes spent second-guessing the caption. Because nothing—absolutely nothing—says *off the grid* like broadcasting it to the entire internet.

The reality is, we're all guilty of it to some degree. We document our escapes, our quiet moments, our breakups, our breakthroughs. We want proof that we lived, and social media provides it. Maybe one day we'll just enjoy a sunset instead of trying to make it go viral. But let's be real—if we do, we'll probably have to tweet about it. And then check five minutes later to see if anyone liked it. And if they didn't, well, did the sunset even really happen? But let's be real—if we do, we'll probably have to tweet about it. But until then, expect to see plenty of beautifully curated posts reminding you just how much someone doesn't care about social media anymore.

9

The Cult of Nostalgia

When Loving Retro Goes Too Far

NOSTALGIA IS A FUNNY thing—the selective amnesia that convinces us everything was better back when internet connections screamed in agony, and watching a movie meant hoping the last renter had the decency to rewind the VHS tape. But sometimes, that innocent longing for the past transforms into something more... aggressive. What starts as an appreciation for vintage aesthetics can quickly morph into a full-blown crusade against anything modern. Before you know it, you're in a heated argument over whether digital music even qualifies as "real music" or if using a Kindle is akin to burning down a public library.

For some, nostalgia isn't just a comforting escape—it's a full-blown ideological movement, like a political party that replaces policy debates with impassioned discussions on why cassette tapes "just hit different." These are the folks who treat smartphones with the same level of disdain one might reserve for gas station sushi. They glare at an e-reader like it personally insulted their grandmother's bookshelf. And nowhere is this phenomenon more visible than in the cutthroat world of typewriters and vinyl records.

ANALOG REBELLION

Inside Typewriter Feuds & Vinyl Elitism

If you thought modern internet debates were heated, try stepping into a room full of typewriter collectors. Few things in life are more ferocious than a debate between two typewriter aficionados, each convinced their machine is the mechanical equivalent of a Shakespearean first folio. Bring up electric typewriters, and someone will need smelling salts. Don't even get them started on "poser" collectors who buy typewriters as décor rather than for their *intended purpose*—which, apparently, is composing letters as if they're Hemingway in a smoke-filled Parisian café.

Ownership alone isn't enough for these purists. They'll scoff if you haven't meticulously restored your typewriter using period-accurate oil and a cleaning technique passed down through generations of grumpy office clerks. If you dare suggest that modern mechanical keyboards replicate the same tactile satisfaction, you'll be banished from the typewriter underground, a place where analog gatekeepers debate whether ribbon ink should be artisanal.

Then there's the world of vinyl records, where the self-proclaimed "true music lovers" reside—those who passionately argue that listening to an album on a digital platform is akin to eating a gourmet meal microwaved in a Styrofoam container. Never mind that half of them own Bluetooth turntables, which is basically like insisting on using a quill pen while secretly appreciating spell-check. Suggest that digital formats might have *some* advantages—portability, affordability, not requiring an entire bookcase to store—and you'll be met with a look of sheer disgust, as if you just suggested replacing all literature with audiobooks read by robots.

These vinyl purists don't just collect records; they wage holy wars over the proper way to clean them. If you dare to touch a record incorrectly, expect a lecture about how the oils from your fingertips will "corrupt the integrity of the sound." And don't even think about admitting that you sometimes enjoy the convenience of digital streaming—unless you're ready for a dramatic monologue on the "soulless compression of modern audio."

ANALOG REBELLION

The Hierarchy of Nostalgia: Who Loved It First?

What's fascinating is how nostalgia, when taken too far, creates invisible hierarchies. Much like coffee snobs debating over single-origin beans and pour-over techniques or cinephiles insisting that anything shot on digital is a crime against cinema, nostalgia enthusiasts develop their own complex pecking orders. It's no longer just about enjoying the past; it's about proving that your version of the past is the *right* one. Enthusiasts don't just love analog—they weaponize it. Suddenly, there are "rules" about what qualifies as "true appreciation" versus "trendy bandwagoning." There are litmus tests for who's a real collector and who's just a passing tourist in the world of nostalgia.

For example, in the world of film photography, enthusiasts will judge your choice of camera like sommeliers judging cheap wine. A vintage Leica? Respectable. A disposable

Kodak? Laughable. And heaven help you if you dare admit that you sometimes *gasp* shoot in digital. In certain circles, this is akin to admitting you enjoy instant coffee in a room full of espresso purists.

Meanwhile, in the world of retro gaming, the hierarchy is even more unforgiving. Emulating classic games on your PC? That's cheating. Owning an original console but using modern HDMI adapters? Barely acceptable. Only those who own a CRT television from the '90s and insist on blowing into their cartridges for "authenticity" receive the highest level of reverence.

ⒶⓃⒶⓁⓄⒼ ⓇⒺⒷⒺⓁⓁⒾⓄⓃ

The Irony of the Analog Purist

The ultimate irony? These champions of the "pure experience" spend hours debating the sanctity of analog media... on Twitter, while scrolling for the best deals on vintage cassette players on eBay. They take to Reddit forums to argue the legitimacy of digital restoration techniques while frantically refreshing their feeds for new comments—because nothing proves your commitment to analog purity quite like using an algorithm to argue against algorithms. They scoff at the soullessness of modern technology but have notifications enabled for every alert about a limited reissue pressing of *Dark Side of the Moon.*

It turns out that even the most devoted analog purists aren't above the convenience of the digital world when it suits them. After all, nostalgia may be selective, but convenience is universal.

ⒶⓃⒶⓁⓄⒼ ⓇⒺⒷⒺⓁⓁⒾⓄⓃ

Not All Old Tech Was Good (And That's Okay)

Nostalgia is a funny thing. It filters out the frustrating, inconvenient, and downright infuriating parts of old technology, leaving us with a glossy, romanticized version of history. We talk about "the good old days" as if they were a golden age of craftsmanship and simplicity, conveniently forgetting the patience and suffering required to survive them. If modern consumers had to rewind VHS tapes, wait three hours to download a single song, or navigate a road trip using MapQuest printouts, they'd stage a full-scale revolt. For every "They don't make 'em like they used to" claim, there's an inconvenient truth we choose to forget—like the absolute *nightmare* that was rewinding VHS tapes.

ⒶⓃⒶⓁⓄⒼ ⓇⒺⒷⒺⓁⓁⒾⓄⓃ

The Horror of VHS and Other Analog Agony

Ah yes, the VHS tape. A relic of home entertainment that many now display proudly on shelves, despite the fact that watching a movie involved a tedious ritual of rewinding, fast-forwarding, and hoping the tape didn't get eaten by the VCR. Nothing quite matched the suspense of renting a movie, popping it into the player, and realizing the last person left it at the credits. Cue the slow, whining screech of rewinding at a speed best described as "painfully inadequate." And if you lost the remote? That meant pressing buttons *on the machine* like some sort of caveman.

And let's not even get started on recording TV shows. If you wanted to tape your favorite program, you had to set the timer correctly and pray that no one changed the channel mid-recording. Otherwise, your *Friends* finale was going to have an unexpected intermission of your dad flipping to the news halfway through.

ⒶⓃⒶⓁⓄⒼ ⓇⒺⒷⒺⓁⓁⒾⓄⓃ

The Nightmare of Dial-Up and Early Internet

People wax poetic about "the early days of the internet," but let's not forget that logging on in the '90s meant enduring the unholy screech of a dial-up modem that sounded like a fax machine being attacked by demons. The joy of "going online" was immediately tempered by the anxiety that someone might pick up the house phone and boot you offline. Yes, kids, there was a time when you had to choose between making a phone call and browsing the web—*and the web loaded one pixel at a time.*

Click on an image, go make a sandwich, and by the time you came back, maybe half the picture had loaded—if your connection didn't randomly drop in the meantime. Want to download a single song? Hope you had a few hours to kill—because that 5MB file was crawling in at 3 kilobytes per second. Meanwhile, today, we throw a tantrum if our Spotify playlist takes more than half a second to load. And let's not forget the era of Limewire, where every file came with a complimentary virus that would either crash your computer or blast static at full volume.

ANALOG REBELLION

The Struggle of Pre-Smartphone Communication

There aren't many people clamoring for a full return to the days of payphones and T9 texting—though, to be fair, some have embraced elements of these relics for the sake of digital detoxing. But the idea of a "simpler time" tends to ignore just how much effort basic communication required. If you weren't home when someone called, they left a voicemail—on your *actual answering machine.* If you lost their number? Tough luck. You better hope you memorized it or had it written down in a tattered address book. And spontaneous plans? Forget it—making plans involved multiple calls, a degree of trust, and a willingness to wait around, hoping your friend actually showed up at the meeting spot.

And let's take a moment to remember T9 texting. For those who weren't around for the struggle, T9 was a predictive text system that required pressing number keys multiple times to get a single letter. Texting "Hey, what's up?" felt like defusing a bomb, as you carefully tapped out each word, hoping you didn't accidentally type "Hex, whys up?" and confuse everyone involved. And God forbid you had to type something with a lot of

S's—because that meant pressing the '7' key four times *for every single S.* If you had to text "Mississippi," you might as well have canceled your plans for the day.

ⒶⓃⒶⓁⓄⒼ ⓇⒺⒷⒺⓁⓁⒾⓄⓃ

Typewriters: Beautiful, Yet Completely Unforgiving

Let's talk about typewriters, the darling of nostalgia-loving writers who claim that "nothing compares to the sound of a real typewriter." Sure, the *clack-clack* of the keys is satisfying, but do you know what's not? Making a typo and realizing you now have three equally terrible choices: 1) Start the entire page over. 2) Use correction tape and accept that your document now looks like it survived a paper war. 3) Pretend the mistake adds "character" and hope no one notices.

And let's not forget that typewriters *don't have delete keys.* One slip-up and your perfectly crafted letter now looks like it was typed by someone in the middle of a nervous breakdown. If you think auto-correct is annoying, imagine a world where every single mistake was permanent. No frantic backspacing, no convenient squiggly red lines—just the cold, unflinching reality of an ink-stamped typo staring back at you forever. Modern writers will obsess over rewording a sentence six times before hitting 'send,' but on a typewriter, your first draft is your final draft, whether you like it or not. Romantic, right?

ⒶⓃⒶⓁⓄⒼ ⓇⒺⒷⒺⓁⓁⒾⓄⓃ

Why Nostalgia Can Be Selective (And That's Okay Too)

The point is, not everything old is superior. Some technology has been rightfully left in the past, and that's okay. Progress isn't the enemy; it's the reason we no longer have to untangle 12 feet of phone cord just to have a private conversation. It's why we don't have to fight with a cassette tape that got chewed up by the car stereo. It's why we don't have to print out MapQuest directions and hope we don't take a wrong turn because there *is no recalculating option* in the real world.

Loving retro aesthetics is fine—owning a record player, using a typewriter, or taking film photos can be enjoyable hobbies. But let's not pretend the past was a technological utopia. It was full of inefficiencies, frustrations, and time-wasting processes that modern technology has (mercifully) improved. The next time someone insists that "things were just *better* back in the day," feel free to remind them of VHS tapes, dial-up internet, and the agony of trying to text "See you soon" on a flip phone. And then hand them a rotary phone and ask them to dial a number without messing up—just for fun. Watch as they hesitate, spin the dial, and realize that if they make a mistake, they have to start all over again. By the time they finish dialing, you could've sent three texts, scheduled a Zoom call, and written an entire novel on your smartphone.

Things You've Romanticized That Actually Sucked

- **Scratchy CDs that skip on your favorite song.** Nothing like gearing up for the chorus only to have your CD player turn it into an avant-garde remix of static and regret.
- **Calling someone's house and hoping they pick up.** The suspense of waiting while the phone rang—would you get them, their mom, or worse, their dad who immediately wanted to know what this was regarding?
- **Waiting three days for film photos that turned out blurry.** Ah yes, the anticipation, the excitement… the crushing disappointment of realizing every single shot looked like it was taken mid-earthquake.
- **Manually rolling down car windows.** Your arm workout for the day, courtesy of 1980s automotive design. And heaven help you if you were in the passenger seat and the driver refused to reach across.
- **Blowing into a Nintendo cartridge to "fix" it.** Did it actually work? Who knows. But we all did it like some kind of ancient ritual passed down through generations.

- **TV static that made you think the channel was haunted.** Late at night, flipping through channels, only to be greeted by a snowy screen and eerie crackling noises that made you reconsider every horror movie you'd ever seen.

- **Printing out MapQuest directions and hoping you didn't take a wrong turn.** There was no "recalculating"—only shame, confusion, and a frantic search for a gas station where someone might know where the heck you were.

- **Getting lost because your friend swore they "knew a shortcut."** Spoiler: They did not.

- **Trying to record a song off the radio without the DJ talking over the intro.** The impossible dream of every mixtape maker. You'd get 90% of the song before some guy interrupted to remind you of the station's upcoming contest giveaway.

- **Waiting for your favorite show to come on instead of streaming it.** Missed it? Too bad. Your only hope was a rerun—*someday*.

- **Rewinding rental tapes and getting charged a fee if you didn't.** Be kind, rewind—or else Blockbuster would hit you with a 50-cent fine and a silent dose of judgment.

- **Carrying a Discman and trying to walk without skipping the track.** Portable, yes. Functional? Only if you walked as if balancing a book on your head.

- **Getting a busy signal instead of voicemail.** Calling over and over, wondering if they were on another call, ignoring you, or had simply vanished into the ether.

- **Accidentally taping over something important.** One wrong button press and *poof*, your sister's graduation ceremony was now a rerun of *Friends*.

10

The Myth of Expensive Analog Living

No Need for a Second Mortgage

Let's be honest—analog can sometimes feel like a hipster status symbol. The $500 film cameras, artisanal leather-bound notebooks, and high-end mechanical watches can make it seem like analog living is reserved for those with deep pockets. But here's the truth: analog isn't about luxury, it's about intention. You don't need to spend a fortune to disconnect from the digital world and enjoy the benefits of a slower, more mindful way of living.

Analog living isn't about chasing expensive objects—it's about making deliberate choices. Whether it's picking up a thrift-store typewriter, finding joy in dollar-store notebooks, or rediscovering the satisfaction of making something by hand, there are plenty of ways to embrace analog without wrecking your budget. Let's explore how.

ANALOG REBELLION

Budget-Friendly Analog Alternatives

1. Writing and Note-Taking

- **Instead of:** A $50 Moleskine journal
  Try: A classic composition notebook or a pocket notebook from a discount store ($1-$5).
  Bonus Tip: Bullet journaling isn't about the brand—it's about the method. Any notebook will work!

- **Instead of:** A $200 Montblanc fountain pen
  Try: A $5 Pilot Varsity or a pack of Bic ballpoints ($2-$10).
  Bonus Tip: Thrift stores and estate sales sometimes have vintage fountain pens that can be restored for cheap.

- **Instead of:** A $25 pack of luxury stationery
  Try: DIY postcards from recycled paper or index cards ($1-$5).

2. Reading and Learning

- **Instead of:** Buying brand-new hardcover books ($30+ each)
  Try: Public libraries (free!), used bookstores, or thrift stores (often $1-$5 per book).
  Bonus Tip: Library book sales often have classics and hidden gems for pennies.

- **Instead of:** An expensive e-reader or audiobooks subscription
  Try: Paperback swap groups, borrowing from friends, or Project Gutenberg for free classic literature.

- **Instead of:** Online courses with hefty price tags
  Try: Free educational resources like local library workshops, YouTube tutorials, and open-access courses.

3. Music and Audio

- **Instead of:** A $300 turntable with $40 vinyl records
  Try: A secondhand record player from Craigslist, thrift stores, or garage sales ($20-$50).
  Bonus Tip: Many record stores have $1 bargain bins full of hidden treasures.

- **Instead of:** A $100 vintage cassette Walkman
  Try: A used CD player and thrift-store CDs ($5-$20 for both).

- **Instead of:** High-end noise-canceling headphones
  Try: Affordable wired earbuds or thrifted speakers that still offer great sound quality.

4. Photography and Creativity

- **Instead of:** A $500 Leica film camera
  Try: A thrift store or garage sale 35mm film camera ($20-$50).
  Bonus Tip: Disposable cameras are a fun, cheap way to try film photography before investing.

- **Instead of:** Fancy art supplies from designer stationery stores
  Try: Basic sketchbooks, dollar-store paints, and thrifted art materials (often

under $10).

- **Instead of:** Expensive calligraphy sets
  Try: Learning with a simple dip pen and ink ($5-$15).

5. Communication and Organization

- **Instead of:** A $300 leather planner
  Try: A simple wall calendar or a pocket planner from the dollar store ($1-$10).

- **Instead of:** Sending digital messages through a paid service
  Try: Writing postcards or letters—stamps are still affordable, and handwritten notes have a personal touch that no text can match.

- **Instead of:** Expensive fountain pen-friendly paper
  Try: Legal pads, printer paper, or even old-school composition books.

6. Hobbies and Hands-On Activities

- **Instead of:** High-end woodworking tools
  Try: Basic hand tools from estate sales or garage sales.
  Bonus Tip: Many community centers have shared workshop spaces where you can use tools for free or a low fee.

- **Instead of:** Designer board games
  Try: Classic games from thrift stores, or DIY your own board game with index cards and creativity!

- **Instead of:** A gym membership
  Try: Walking, jogging, or DIY fitness routines with free weights or resistance bands.

The Beauty of Low-Tech Simplicity

The real magic of analog isn't in the price tag—it's in the experience. A cheap notebook filled with your thoughts is more valuable than an expensive one left blank. A $5 thrift-store film camera can capture moments just as beautifully as a $1,000 one. The point isn't perfection—it's presence.

By choosing budget-friendly analog alternatives, you can reclaim a slower, more intentional way of living without spending a fortune. Analog living is about creativity, adaptability, and rediscovering what truly matters—not about chasing expensive, trendy products.

Part 3

HOW TO SURVIVE & THRIVE IN THE ANALOG-DIGITAL WORLD

11

The Analog Cure

How Going Analog Boosts Mental Health

Why Our Brains Need Analog

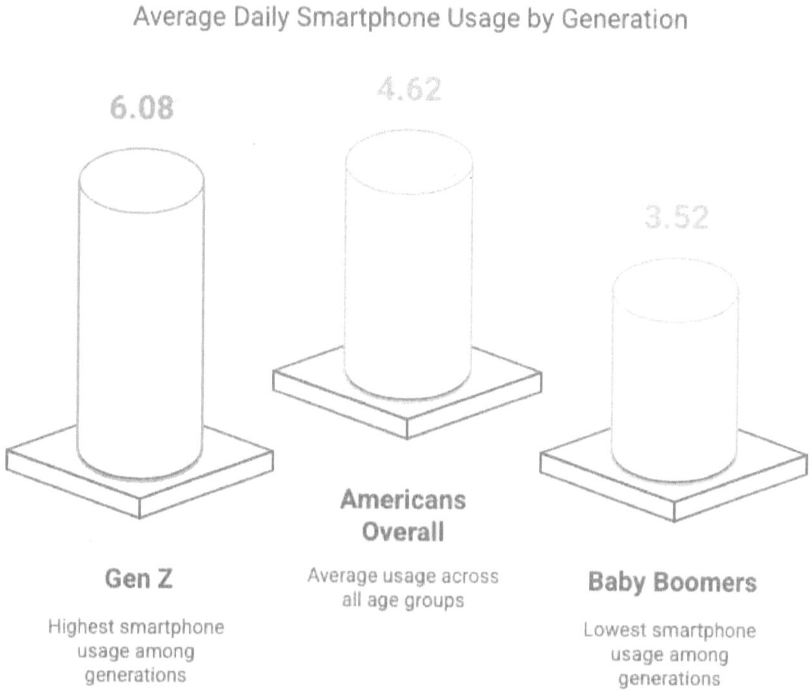

In a world increasingly dominated by digital devices, a 2023 study by Harmony Healthcare IT[1] revealed that Americans spend an average of 4 hours and 37 minutes per day on their phones, with screen time varying significantly by generation. Gen Z leads the way, spending 6 hours and 5 minutes daily, while Baby Boomers average 3 hours and 31 minutes. This reliance on smartphones has grown, with 45% of Americans admitting they feel addicted to their devices. The study also found that 40% of respondents are actively trying to reduce their screen time, citing reasons such as improving mental health,

---

1. Kaylin. "Phone Screen Time Addiction- New Survey Data & Statistics." *Healthcare Data Management Software & Services | Harmony Healthcare IT*, 8 Jan. 2024, www.harmonyhit.com/phone-screen-time-statistics/.

increasing focus, and enhancing overall well-being. Despite these efforts, the pervasive use of phones for entertainment, boredom, and distraction continues to shape daily habits, highlighting the challenges of balancing digital engagement with mental and physical health.

ANALOG REBELLION

The Science of Handwriting and Memory Retention

If you've ever written something by hand and remembered it better than a typed note, you're not imagining things. Studies show that handwriting engages the brain in ways that typing simply does not. Researchers at Princeton and UCLA[2] found that students who took handwritten notes retained more information and understood concepts better than those who typed their notes on a laptop. This is because handwriting requires deeper cognitive processing, while typing often leads to verbatim transcription without true engagement with the material. Writing by hand forces us to process information more deeply, leading to stronger comprehension and recall.

Handwriting also activates neural pathways associated with learning and creativity. The act of forming letters and words manually requires greater cognitive effort, which strengthens memory encoding. In other words, putting pen to paper helps cement ideas in your brain far more effectively than hammering away at a keyboard.

ANALOG REBELLION

The Power of Reading Physical Books

2. Mueller, Pam, and Daniel Oppenheimer. *Psychological Science OnlineFirst*. 22 May 2014.

E-readers and audiobooks have their place, but when it comes to deep reading and comprehension, physical books still reign supreme. Studies have shown that people who read on paper absorb more information than those who read on screens. The tactile nature of flipping pages, the ability to highlight and annotate physically, and the lack of screen glare all contribute to improved retention.

A study[3] found that Kindle users had a significantly harder time recalling plot details than those who read a physical book. Researchers suggest that the sensory experience of holding a book and visually tracking progress through physical pages creates a stronger connection between the reader and the material. Physical books also encourage deeper focus—without notifications or the temptation to switch apps, reading becomes an immersive experience rather than a fragmented one. Digital screens, on the other hand, promote skimming rather than deep reading, as the constant availability of hyperlinks, ads, and background applications encourages multitasking and shorter attention spans. Studies have found that prolonged screen exposure can reduce cognitive engagement, making it harder to retain information and fully absorb complex narratives.

ANALOG REBELLION

The Calming Effects of Physical Journaling

Journaling is one of the simplest and most effective ways to clear the mind, reduce stress, and enhance self-awareness. I once found myself overwhelmed by a flood of thoughts late at night, unable to sleep. Instead of endlessly scrolling through my phone, I grabbed a notebook and started writing. What started as a jumbled mess of frustrations slowly transformed into a sense of clarity. By the time I closed the journal, my mind felt lighter, and I finally got the deep sleep I had been chasing. This small act of putting pen to paper had an immediate, calming effect—something no app had ever achieved for me.

---

3. Mangen, Anne, et al. "Comparing Comprehension of a Long Text Read in Print Book and on Kindle: Where in the Text and When in the Story?" *Frontiers in Psychology*, vol. 10, no. 38, 15 Feb. 2019, https://doi.org/10.3389/fpsyg.2019.00038.

But the medium matters. Writing in a physical journal has been shown to provide greater emotional relief than typing on a screen. The act of writing longhand slows down the thought process, allowing for deeper reflection and emotional processing.

A study published in the journal *Advances in Psychiatric Treatment*[4] found that expressive writing—writing about thoughts and feelings without judgment—significantly reduces stress and anxiety. When done on paper, this practice encourages mindfulness, helping individuals stay present and engaged rather than distracted by the constant influx of digital stimuli.

Furthermore, journaling on paper eliminates the distractions of the digital world. There are no pop-ups, no autocorrect, and no algorithm steering your thoughts. It's just you, your thoughts, and the slow, deliberate process of writing—something that has been a therapeutic practice for centuries.

ANALOG REBELLION

Analog as a Form of Mindfulness

Mindfulness is about being present in the moment, and analog activities naturally cultivate this state. Whether it's sketching, knitting, woodworking, or simply writing a letter, these activities engage the mind fully, allowing it to slow down and reset. When we disconnect from screens and focus on a tactile experience, we engage with the world in a way that is both grounding and restorative.

---

4. "Speaking of Psychology: Expressive Writing Can Help Your Mental Health, with James Pennebaker, PhD." *Apa.org*, 2024, www.apa.org/news/podcasts/speaking-of-psychology/expressive-writing.

Research on mindfulness practices[5] has repeatedly shown that slowing down and engaging in intentional, hands-on activities lowers cortisol levels (the stress hormone), reduces anxiety, and improves overall well-being. Analog activities provide a natural way to integrate mindfulness into daily life without needing a meditation app or a guided exercise.

ANALOG REBELLION

Bringing Analog into Your Daily Routine

So how can you introduce more analog practices into your life to support mental well-being? Here are some simple, effective ways:

- **Start a handwritten journal** – Write about your thoughts, reflections, or even daily events to enhance clarity and emotional processing.
- **Read a physical book** – Set aside time to read without distractions and experience the joy of deep focus.
- **Send handwritten letters or postcards** – Strengthen relationships while enjoying the satisfaction of a tangible, personal connection.
- **Take notes by hand** – Whether it's for work, study, or brainstorming, handwritten notes improve retention and engagement.
- **Engage in tactile hobbies** – Painting, gardening, and woodworking encourage mindfulness and creativity.

---

5. Altschul, Drew M., and Ian J. Deary. "Playing Analog Games Is Associated with Reduced Declines in Cognitive Function: A 68-Year Longitudinal Cohort Study." *The Journals of Gerontology: Series B*, vol. 75, no. 3, 14 Feb. 2020, pp. 474–482, academic.oup.com/psychsocgerontology/article/75/3/474/5628188, https://doi.org/10.1093/geronb/gbz149.

By embracing these analog habits, you create moments of stillness and focus in a world that constantly pulls you in multiple directions. The analog world is not just about nostalgia—it's about reclaiming mental clarity, reducing stress, and fostering a more intentional way of living.

So pick up a pen, open a book, or start a new hands-on hobby—your brain will thank you. Take the first step toward reclaiming your attention, one analog moment at a time, and experience the depth and clarity that digital distractions can never replace.

12

The Analog Starter Pack

How to Reclaim Your Attention Span

Dumbphone Weekends: How to Do It Without Panic

So, you want to escape the digital leash for a weekend? Bold move, my friend. Stepping away from your smartphone is like taking off a pair of too-tight shoes—uncomfortable at first, but ultimately a huge relief. Here's how to survive a dumbphone weekend without hyperventilating into a paper bag.

Step 1: Acknowledge Your Addiction

First, admit it: you check your phone more often than a nervous contestant on a game show. It's not your fault—these devices are engineered to hijack your dopamine system. But knowing this is power. It's time to take back control. Recognizing that you've been trapped in a loop of mindless scrolling is the first step toward breaking free.

Step 2: Choose Your Weapon

You don't need to go full rotary dial (unless you really want to). A simple flip phone or an old-school Nokia brick will do the trick. These things are built like tanks and have batteries that last longer than your willpower at a donut shop. If you really want to go old school, consider a landline for the weekend—yes, they still exist.

Step 3: The Digital Detox Prep

Before you unplug, tell people you're going off-grid. Some might panic at the thought of not reaching you instantly, so reassure them by setting expectations upfront. Let them know when you'll be back online and offer an alternative way to reach you in case of true emergencies. Otherwise, your friends might assume you've been abducted. Set up an emergency contact method (yes, your landline still works), print out any maps you might need, and jot down important phone numbers—because memorizing numbers went extinct with the dial-up modem. Also, prepare for the unexpected: write down your schedule, make a plan for entertainment, and brace yourself for the empty moments.

Step 4: Surviving the Withdrawal

The first few hours will be rough. You'll instinctively reach for your pocket, only to grasp air. You might experience phantom vibrations, as if your leg itself is sending you notifications. This is normal. Breathe through it. Replace your scrolling habit with something tangible—flip through a magazine, doodle, or engage in that ancient art form called 'talking to people in person.' Walk outside, watch the sky, rediscover the texture of the real world. It will be unfamiliar at first, but your brain will start reorienting itself to the present.

ANALOG REBELLION

Comparison Table: Analog vs. Digital Lifestyle

Aspect	Digital Lifestyle	Analog Lifestyle
Distractions	Constant notifications, pop-ups, and alerts	Minimal interruptions, focused attention
Productivity	Easily sidetracked by emails, social media, and messages	Deep focus, fewer distractions
Social Interaction	More online conversations, fewer face-to-face moments	Encourages in-person communication
Memory Retention	Quick searches replace actual knowledge storage	Writing by hand improves memory and cognition
Creativity	Algorithm-driven content consumption	More organic, intentional creativity
Emotional Health	Doomscrolling, FOMO, anxiety	Lower stress, improved mindfulness
Sleep Quality	Blue light disrupts melatonin production	No screen interference, deeper rest

ANALOG REBELLION

Common Challenges & How to Overcome Them

- **FOMO (Fear of Missing Out):** Your brain will try to convince you that you're missing something important. You're not. If anything, you're gaining a break from information overload.
- **Boredom:** Your phone has been your go-to boredom killer. Have a plan—bring a book, start a journal, or go for a walk.
- **Work Demands:** If work expects you to be on call 24/7, set clear boundaries ahead of time. Let your team know when you'll be available again and provide an emergency contact method.
- **Urge to "Just Check One Thing":** Leave your phone in a different room. If you must check something, give yourself a strict time limit.
- **Navigational Challenges:** If you're going somewhere unfamiliar, print a map or write down directions before heading out.

Step 5: Revel in the Freedom

By day two, something magical happens: you stop missing your phone. You start noticing details you previously ignored—the way the morning light filters through your window, the texture of a book cover, the subtle background sounds of the world around you. You might find yourself engaging in hobbies you forgot you enjoyed, like sketching, journaling, or cooking a meal without rushing through it. Conversations become deeper, walks feel more immersive, and your thoughts have the space to wander freely. You feel lighter, more present. Food tastes better. Birds sing louder. You realize how much of your life was spent staring at a glowing rectangle. Congratulations, you've temporarily escaped the algorithm's grip. You might even find yourself reaching for a book instead of your usual scrolling routine. Creativity starts to flow, and your attention span begins to recover.

Step 6: Reintegrate With Caution

Monday morning will arrive, and you'll have to return to the digital world. But now, you've seen the light. Consider keeping the dumbphone for future weekends or setting boundaries like "No phone after 8 PM." You've proven you can survive without it—why not make this a habit? You may even find yourself extending these breaks into daily life, cutting back on social media, and redefining how much of your attention belongs to you rather than an app.

ANALOG REBELLION

Essential Analog Tools That Aren't Just Hipster Bait

Going analog doesn't mean embracing technology from the Stone Age; it means using tools that enhance your life rather than siphon your attention. For example, a paper planner allows for better long-term goal tracking without the distractions of notifications and ads. A physical book lets you immerse yourself in a story without the temptation of multitasking, unlike an e-reader with built-in apps. Here's a list of analog essentials that are actually useful, not just Instagram props for guys with waxed mustaches.

Recommended Resources: Further Reading & Exploration

If you want to dive deeper into the benefits of analog living, here are some excellent resources:

- **Books:**
    - *Digital Minimalism* by Cal Newport
    - *The Revenge of Analog* by David Sax
    - *How to Break Up with Your Phone* by Catherine Price
    - *The Shallows* by Nicholas Carr

- **Documentaries:**
  - *The Social Dilemma* (Netflix) – Explores the dangers of social media and digital addiction
  - *Minimalism: A Documentary About the Important Things* (Netflix) – Examines how simplifying life improves focus and well-being
- **Websites & Articles:**
  - The Analog Office – A blog dedicated to analog tools for productivity
  - Cal Newport's Blog – Deep work, focus, and digital minimalism insights
  - The Digital Declutter Experiment – A practical guide to reducing screen time

By exploring these resources, you can continue your journey toward reclaiming focus, creativity, and a life less dictated by screens. These books provide practical strategies for reducing digital dependency, while the documentaries offer eye-opening insights into the effects of excessive screen time. The recommended websites and articles serve as ongoing guides to maintaining a more intentional, analog-centered lifestyle. Whether you start with a small change or go all in, each of these resources can help you build a sustainable approach to digital minimalism.

Final Thoughts: Analog for Sanity, Not Aesthetic

Going analog isn't about looking cool—it's about reclaiming your attention, your time, and your ability to focus. Whether it's a weekend detox or a full lifestyle shift, these tools will help you step away from the endless digital barrage and reconnect with reality. And who knows? You might actually enjoy the silence. More than that, you might just find your creativity, patience, and joy returning in ways you never expected.

ⒶⓃⒶⓁⓄⒼ ⓇⒺⒷⒺⓁⓁⒾⓄⓃ

The Analog Survival Kit

Surviving the analog life isn't just about ditching screens—it's about embracing the weird, wonderful, and sometimes hilariously impractical reality of a world without instant gratification.

Here's your essential Analog Survival Kit:

1. **Patience** – Because analog is slow, and you will have to wait.
2. **A Lamp** – Books don't glow in the dark, and neither do newspapers.
3. **A Wristwatch** – Because pulling out your phone for the time is how the digital trap starts.
4. **A Rolodex** – So you can pretend it's the '80s while flipping through business cards.
5. **A Notebook & Pen** – The original Notes app, only crash-proof.
6. **A Real Map** – GPS is for quitters.
7. **A Pocket Calendar** – Because scheduling without push notifications builds character.
8. **Cash Money** – Ever tried paying with a credit card in a cash-only café? Painful.
9. **Stamps & Envelopes** – The original 'send' button, just slower and more adhesive.
10. **A Magnifying Glass** – Useful for fine print and feeling like Sherlock Holmes.
11. **A Newspaper** – Doubles as reading material and an emergency rain hat.
12. **A Vinyl Record Player** – Because music should come with a workout (lifting LPs is no joke).

13. **A Deck of Cards** – For when your Wi-Fi 'accidentally' stops working.
14. **An Old-School Alarm Clock** – So your phone doesn't wake you up and suck you in.
15. **A Mechanical Pencil** – The closest thing to Ctrl+Z in the analog world.
16. **A Filing Cabinet** – Where important papers go to be forgotten.
17. **A Library Card** – The original streaming service, with better recommendations.
18. **A Flashlight** – Because sometimes you forget to charge your lamp.
19. **A Landline Phone** – The most radical way to screen calls.
20. **A Rock** – No real reason. Just... a rock. Good for paperweights, stress relief, and occasional self-defense.

13

The Joy of Friction

Why Harder is Sometimes Better

Why Effort Makes Life More Meaningful

If everything in life came easily, we'd all be miserable. Think about it: when was the last time you felt truly proud of something that required zero effort? The reality is, struggle breeds satisfaction. It's why climbing a mountain feels better than taking an escalator and why finishing a difficult book is more rewarding than binge-watching a show where nothing happens.

Effort forces us to engage fully with the world. When you put in the work, whether it's learning a new skill, cooking from scratch, or even just writing something by hand instead of typing, you create a deeper connection to the experience. Convenience may be seductive, but meaning is found in the moments where we push through discomfort and achieve something real. Think about learning to ride a bike. As a kid, you start with training wheels, wobbling down the sidewalk. Then comes the dreaded moment when they come off, and suddenly, you're flailing, scraping knees, and questioning your life choices. But when you finally pedal without falling, it's not just about balance—it's about independence, confidence, and the sheer joy of knowing you conquered something difficult.

Think about hobbies—playing an instrument, gardening, woodworking, or even perfecting a sourdough starter. These things take time and patience, but that's exactly what makes them so satisfying. The best things in life are those that demand our attention, skill, and dedication. Mastering a skill, even one as small as making the perfect omelet, is rewarding precisely because it wasn't instantaneous.

Challenges create memories, and memories shape identity. No one reminisces about the effortless things in life. Instead, we cherish the time we struggled to put together a puzzle as a kid or the feeling of triumph after finally nailing that tricky yoga pose. Effort leaves an imprint in our minds, making moments richer and more valuable.

I remember trying to build a treehouse with my dad when I was ten. We had no real blueprint—just a vague idea and a pile of wood. Every nail I hammered in was slightly crooked, and the floorboards were uneven, but when we finally climbed up and sat on our wobbly masterpiece, it felt like the greatest accomplishment of my childhood. That treehouse, flawed as it was, stood as a testament to the joy of effort. And it still stands—barely. I half expect an archaeological team to mistake it for an ancient ruin one day.

ANALOG REBELLION

The Hidden Satisfaction of the Hard Way

Take something as mundane as fixing a leaky faucet. You could call a plumber and have it done in minutes, or you could wrestle with wrenches and washers, swearing under your breath the entire time. But when you finally stop the drip, there's an undeniable rush of accomplishment. The challenge itself makes the end result feel more valuable.

My friend once spent three weeks assembling a bookshelf from scratch, refusing to buy a pre-made one. Every evening after work, he'd sand, measure, and meticulously stain each piece. By the time he was done, the shelf wasn't perfect—some parts didn't quite line up—but he loved it more than anything he could have bought in a store. Why? Because every inch of it reflected his effort and patience. Plus, now he gets to tell guests, "Yes, I built that," in a tone of exaggerated self-importance.

ANALOG REBELLION

The Surprising Science of Doing Things the Hard Way

It turns out, science is on the side of friction. Studies show that when we put in more effort, our brains actually value the outcome more. Psychologists call this the "IKEA effect"[1] —the idea that we love things more when we've had a hand in making them, even

1. Norton, Michael I., et al. "The IKEA Effect: When Labor Leads to Love." *Journal of Consumer Psychology*, vol. 22, no. 3, July 2012, pp. 453–460, https://doi.org/10.1016/j.jcps.2011.08.002.

if they turn out slightly crooked. (Yes, even that wobbly bookshelf you spent four hours assembling. At least it hasn't collapsed—yet.)

Handwriting notes instead of typing improves memory retention. Cooking a meal from scratch strengthens your relationship with food and your ability to survive an apocalypse without Uber Eats. Even walking instead of taking the elevator can lead to better health and a stronger sense of autonomy.

The "Desirable Difficulty"[2] principle suggests that struggling through a task enhances learning. Research in cognitive psychology has shown that when students put in more effort—like solving a problem before being given the answer—they retain the knowledge better. It's why things that come too easily are often forgotten, while the hard-earned lessons stick with us for years. Just like you'll always remember the time you tried to assemble IKEA furniture without reading the instructions. It's a rite of passage, much like learning a new language—frustrating, full of mistakes, and at some point, you probably want to throw something across the room. But when it finally clicks, whether it's fluently ordering a coffee in Paris or standing triumphantly in front of a semi-stable bookshelf, the effort makes the success that much sweeter. Spoiler: it didn't go well.

ANALOG REBELLION

The Flip Side: When Convenience is Actually Good

Of course, not every shortcut is bad. No one is arguing that we should churn butter by hand just for the sake of suffering. Technology and efficiency have their place—life-saving medicine, safe transportation, and instant access to information are all gifts of modern progress. The key is finding balance. While technology makes life easier, incorporating effort into daily activities—like cooking a meal from scratch or writing by hand—can create deeper connections and more fulfilling experiences. It's about knowing when to

---

2. Bjork, Elizabeth L, and Robert Bjork. *Making Things Hard on Yourself, but in a Good Way: Creating Desirable Difficulties to Enhance Learning Using Testing to Improve Learning and Memory*. 2010.

appreciate convenience and when to embrace a little struggle for personal growth. The key is balance: knowing when to embrace ease and when to lean into effort. So, yes, go ahead and enjoy your pre-sliced bread. Just don't let all of life become a drive-thru experience.

The Joy of Mastery Through Effort

Struggle also brings with it a certain level of mastery that shortcuts simply can't provide. When we take the long road to learning or achieving something, we understand it more deeply. A person who builds a bookshelf from scratch has a greater appreciation for craftsmanship than someone who buys one pre-assembled. Likewise, someone who writes a handwritten letter instead of sending a quick text is more invested in the act of communication.

Friction also fosters creativity. Take Vincent van Gogh, for example—he worked with limited colors and thick, unblended brushstrokes, not because he had to, but because constraints forced him to innovate and develop his unique style. Similarly, some of the greatest inventions in history, like the telephone and the airplane, emerged from the challenge of working with limited resources and technology. Have you ever noticed how some of the most creative solutions arise when technology isn't available? When a tool breaks, people have to think outside the box to solve the problem. When an artist is limited to just a few colors, they become more inventive. Constraints—whether time, resources, or energy—force us to be resourceful, and that often leads to a deeper level of satisfaction.

Technology has made many things easier, but at what cost? When was the last time you took the slow route just for the sake of experiencing it? The scenic route on a road trip, the longhand version of a note, the full version of a song instead of the shortened radio edit—these things invite a sense of presence that is otherwise lost in the fast-paced, instant-access world we live in.

ⒶⓃⒶⓁⓄⒼ ⓇⒺⒷⒺⓁⓁⒾⓄⓃ

Why Friction Creates a More Fulfilling Life

Ultimately, effort creates depth. In relationships, it's the effort that strengthens bonds—planning a thoughtful surprise, writing a heartfelt letter, or simply being present and engaged in conversation. In personal growth, challenges push us to evolve. In work, the most rewarding projects are those that stretch our abilities.

The convenience of the digital age is wonderful in many ways, but when we eliminate all friction, we lose something fundamental: the deep satisfaction that comes from overcoming obstacles. When everything is easy, nothing feels valuable.

So the next time you're tempted to take the easy route, consider this: maybe harder is better after all. Life isn't just about efficiency—it's about depth, connection, and the joy of fully engaging with the world around you. Embrace the friction, because sometimes, the best things in life require a little extra effort—just like that wobbly treehouse I built with my dad. It may not have been perfect, but the experience of creating it, the scraped knuckles and misplaced nails, made it infinitely more meaningful. The same goes for life—lean into the effort, and you'll find a satisfaction that shortcuts simply can't offer.

Call to Action: Choose Your Friction

What's one thing you can do this week that adds a little more effort into your life? Maybe it's handwriting a letter to a friend instead of sending a text. Maybe it's taking the stairs instead of the elevator. Maybe it's assembling IKEA furniture *without* losing your sanity. Try one small act of intentional friction and notice how it changes the way you engage with the world.

ANALOG REBELLION

Signs You've Gone Too Far with Analog

1. You train pigeons to deliver messages.

2. You use a map instead of Google Maps.

3. You unironically say, "Back in my day..."
4. You refuse to use auto-correct and insist on rewriting entire paragraphs.
5. Your idea of a playlist is making actual mix tapes.
6. You carry a pocket watch and refer to the time as "a quarter past."
7. You send handwritten letters for texts and wait a week for a reply.
8. You develop an actual relationship with your local librarian.
9. You print out MapQuest directions and still get lost.
10. You own more notebooks than a stationary store but never finish one.
11. You tell your friends, "Let's meet at the café at noon," with no backup communication plan.
12. You refuse to take pictures and insist on "remembering it with your mind."
13. You use a fountain pen... and carry a bottle of ink just in case.
14. You've taught yourself Morse code "just in case the grid goes down."
15. You still burn CDs to make custom playlists.
16. You tell people to "look it up in the dictionary" instead of Googling.
17. You prefer a landline because "it just sounds better."
18. You make your own butter "for the principle of it."
19. You get excited when you find a payphone and actually use it.
20. You proudly declare, "I don't even own a smartphone," and expect applause.

14

Conclusion

The Future is Intentional

If you've made it this far, congratulations—you've survived an entire book without getting distracted by your phone (or at least, not *too* many times). That in itself is proof that your attention span is still in there somewhere, waiting to be reclaimed. And that, my friend, is no small feat in an era where our own brains crave 'just one more scroll.'

We've spent this journey rediscovering the joys of vinyl records, paper planners, landline phones, and even the art of patience. But this isn't just about nostalgia—it's about reclaiming focus, autonomy, and mental well-being in a world that thrives on distraction. The Great Analog Rebellion isn't a rejection of technology—it's a refusal to let it run our lives unchecked. It's about making mindful choices in a world that constantly demands our attention, about resisting the pull of the algorithmic black hole that wants to monetize every second of our existence.

Analog living isn't about throwing your smartphone into a lake (though, if that's your style, I respect the commitment). It's about balance. It's about knowing when digital tools serve us—and when we're serving them. It's about reclaiming boredom as a space for creativity, choosing depth over distraction, and experiencing life without an algorithm filtering it for us. It's about rediscovering the satisfaction of physical effort, the thrill of a challenge, and the quiet joy of something as simple as turning the pages of a well-worn book.

So what now? Start small. Swap out one digital habit for an analog one and see how it feels. Write in a notebook instead of typing. Listen to an album from start to finish without skipping. Use a map (okay, maybe have GPS as backup). Find joy in slowness, in imperfection, in doing things the hard way just for the sake of it. Pick up a pen. Try cooking something entirely from scratch. Carry a book with you instead of doomscrolling while waiting in line.

By choosing inconvenience, you create space for presence. In that resistance to instant gratification, you'll rediscover focus. And in stepping away from the noise, even briefly, you might just hear your own thoughts again.

The digital world isn't going anywhere, but that doesn't mean we have to be at its mercy. The Great Analog Rebellion is just getting started—and now, you're part of it. How will you take part? You don't have to delete all your accounts, throw away your gadgets, or move to a cabin in the woods (unless that's the dream, in which case, go for it). You just have to make choices that serve *you* rather than the machine.

Welcome back to reality. It's nice here, isn't it? Slow, maybe. But nice.

Please Consider Leaving a Review

HELLO THERE!

As an author, I know just how important reviews are for getting the word out about my work. When readers leave a review on Amazon, it helps others discover my book and decide whether it's right for them.

Plus, it gives me valuable feedback on what readers enjoyed and what they didn't.

So if you've read my book and enjoyed it (or even if you didn't!), I would really appreciate it if you took a moment to leave a review on Amazon. It doesn't have to be long or complicated - just a few words about what you thought of the book would be incredibly helpful.

Thank you so much for your support!

Jeff

Also by

Our catalog is constantly growing!

Visit AdultingHardBooks.com

For our other titles and free bonuses!

www.ingramcontent.com/pod-product-compliance
Lightning Source LLC
LaVergne TN
LVHW090411210926
848480LV00002B/532
9798230987642